Designing Organic
Syntheses

Designing Organic Syntheses

Syntheses

**A Programmed Introduction
to the
Synthon Approach**

STUART WARREN

University Chemical Laboratory
Cambridge

JOHN WILEY & SONS
Chichester · New York · Brisbane · Toronto · Singapore

Library of Congress Cataloging in Publication Data:
Warren, Stuart G.
Designing organic synthesis.

1. Chemistry, Organic—Synthesis—Programmed
instruction. I. Title.

QD262.W28   547'.2   77-15479

ISBN 0 471 99612 2

Printed and bound by Antony Rowe Ltd Eastbourne

PREFACE

There are some excellent books written about organic synthesis but they mostly present complete syntheses of complicated molecules. They translate from the language of synthesis into that of organic chemistry. I have tried in this book to teach people to speak the language of synthesis themselves, using the grammar of synthon and disconnection. The programme was originally written for second year university students and re-written after they had used it.

I thank Drs. Ian Fleming, Ted McDonald, Jim Staunton, and Peter Sykes, and the IB advanced chemists of 1976-7 for their perceptive comments, Dr. Denis Marrian for his enthusiastic help in converting the original sketches into a book, and Miss Lesley Rolph for typing the printed words.

This programme is now only part of a comprehensive package on designing organic syntheses; my textbook 'Organic Synthesis : The Disconnection Approach' and its companion 'Workbook' (both published by Wiley in 1983) being the rest. All follow the same approach but use different examples.

Cambridge 1983   Stuart Warren

NOTE FOR INSTRUCTORS

The programme aims to allow students to teach themselves but it shouldn't mean any less work for you. Because the students discover what they don't know, they should have more sensible questions to ask you than if they were reading a textbook or revising from their notes. My aim is to give you more time for real teaching. The programme should do the ground work and you should be able to set suitable problems and discuss then profitably. The programme itself has plenty of problems of this sort (see review and revision problems, and those at the end without worked solutions), and the source books below will give you hundreds more. The literature references are so that you can look up details if you are asked - I imagine few students will use them.

Source Books:

N. Anand, J. S. Bindra and S. Ranganathan,
'Art in Organic Synthesis',
Holden Day, San Francisco, 1970.

J. Ap Simon (editor),
'The Total Synthesis of Natural Products',
Wiley, New York, 1973 - 81, 4 Volumes.

K. Nakanishi et al.,
'Natural Products Chemistry',
Academic Press, New York, 1974, 2 Volumes.

British Pharmacopoeia Commission,
'British Approved Names 1981', HMSO, London,
1981 and later supplements.

CONTENTS

6. Functional Group Addition,
   371-383.
   (a) Strategy of Saturated
       Hydrocarbon Synthesis,
       371-380.
   (b) FGA to Intermediates,
       381-383.
7. Molecules with Unrelated
   Functional Groups,
   384-390.

Though the programme may introduce you to some new reactions, its main aim is to suggest an analytical approach to the design of syntheses. You therefore need to have a reasonable grounding in organic chemistry so that you are familiar with most basic organic reactions and can draw out their mechanisms. If you are a third year university student, a graduate, or someone with experience of organic chemistry in practice you will probably be able to work straight through the programme to learn the approach and not need to learn any new material. If you are a second year university student or someone with a limited knowledge of organic reactions you may find you need to learn some reactions as you go along. I have given references to these books to help you:

'The Carbonyl Programme':
"Chemistry of the Carbonyl Group, A Programmed Approach to Organic Reaction Mechanisms", Stuart Warren, Wiley 1974. This programme leads up to the present one.

'Fleming':
"Selected Organic Syntheses", Ian Fleming, Wiley 1973. Synthesis from the other side: notable examples of organic syntheses carefully explained in detail.

'Tedder':
"Basic Organic Chemistry", J. M. Tedder, A. Nechvatal, and others, Wiley, 5 volumes 1966-1976. A complete textbook of organic chemistry. Explains all the reactions used in the programme and describes many syntheses in detail.

'Norman':
"Principles of Organic Synthesis", R. O. C. Norman, Methuen, 1968: A textbook of organic chemistry from the point of view of synthesis. An excellent source book for all the reactions used in this programme.

Whoever you are, you will certainly find discussion with your fellow students one way to get the most out of the programme and you may well find it is a good idea to work on the more difficult problems together. The review problems, revision problems, and problems without worked solutions are ideal for this. In some cases I have given references to the original literature so that you can find out more details of the various possible approaches for yourself if you want to. It isn't necessary to look up any of these references as you work through the programme.

The point of programmed learning is that
you learn at your own pace and that you yourself check on your own progress. I shall
give you information and ideas in chunks
called frames, each numbered and separated by
a black line. Most frames contain a question,
sometimes followed by a comment or clue, and
always by the answer. You must WRITE DOWN on
a piece of paper your answer to each question.
You'll find that you discover as you do so
whether you really see what is being explained
or not. If you simply say to yourself 'Oh,
I can do that, I don't need to write it down',
and look at the answers, you're missing the
opportunity to check on your own progress as
well as probably deceiving yourself.

When you are ready to start, cover the
first page with a card and pull it down to
reveal the first frame. Read and act on that
frame, then reveal frame 2 and so on. If you
are unfamiliar with the disconnection approach,
I suggest you read the introduction 'Why
bother with disconnections' so that you can
see what I'm driving at. Otherwise the first
sections of the programme may seem rather
pointless.

The aim of this programme is that you
should learn how to design an organic synthesis for yourself. Supposing you wanted to
make this compound:

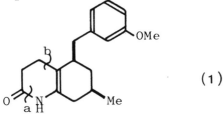

(1)

You would find that it had already been made
by the route outlined on the chart on the next
page. You could then buy the starting materials (compounds 2, 3, 5, 8, and MeI) and set
to work. But supposing 1 had never been
synthesised. How would you design a synthesis for it? You don't know the starting materials - all you know is the structure of the
molecule you want - the TARGET MOLECULE.
Obviously you have to start with this structure and work backwards. The key to the problem is the FUNCTIONAL GROUPS in the target
molecule, in this case the nitrogen atom, the
carbonyl group, the double bond and the benzene ring with its methoxyl group. You should
learn from the programme that for most functional groups there are one or more good
DISCONNECTIONS - that is imaginary processes,
the reverse of real chemical reactions, which
break a bond in the target molecule to give us
the structure of a new compound from which the
target molecule can be made.

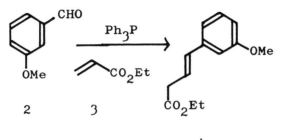

2　　　　3

4

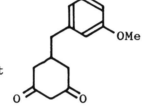

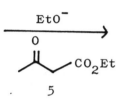

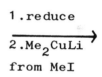

5

6

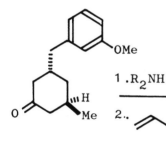

7　　　　8　　　　1

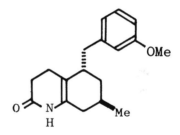

9

Here the first disconnection (a̲) was of a C-N bond, the second (b̲) of a C-C bond taking us back to compounds (7) and (8):

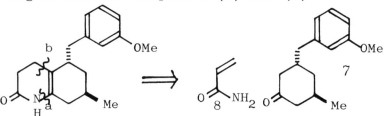

These are in fact standard disconnections which you will meet in sections G and C of the programme. The first part of the programme (Sections B to H) shows you how to use disconnections and which disconnections are good ones. The second part shows you how to choose between alternative series of disconnections to get good synthetic schemes.

When you have finished the programme you should be able to design syntheses for molecules of the complexity of (1). Given this problem, you might not come up with the solution shown in the chart because there is no single "right answer" to a synthesis problem - any given molecule may well be made successfully by several different routes. In practice each of your proposals would have to be tested in the lab., and your overall scheme modified as a result. There were in fact several changes of plan in the synthesis of (1) and you can read more about the details in Stork's article in Pure and Applied Chemistry, (1968, 17, 383) where you will see that he used (1) as an intermediate in the

synthesis of the alkaloid lycopodine (9).
That is a target molecule beyond the scope of
this programme, but organic chemists plan
such syntheses using the same principles as
you will learn here. You must first start at
the beginning and learn in Section A how to
use simple disconnections.

Disconnection: An analytical operation, which breaks a bond and converts a molecule into a possible starting material. The reverse of a chemical reaction. Symbol $\Longrightarrow$ and a curved line drawn through the bond being broken. Called a dislocation by some people.

FGI: Functional Group Interconversion: The operation of writing one functional group for another so that disconnection becomes possible. Again the reverse of a chemical reaction. Symbol $\Longrightarrow$ with FGI written over it.

Reagent: A compound which reacts to give an intermediate in the planned synthesis or to give the target molecule itself. The synthetic equivalent of a synthon.

Synthetic Equivalent: A reagent carrying out the function of a synthon which cannot itself be used, often because it is too unstable.

Synthon: A generalised fragment, usually an ion, produced by a disconnection. (Some people also use synthon for a synthetic equivalent).

Target Molecule: The molecule whose synthesis is being planned. Usually written TM and identified by the frame number.

1. You know that you can make t-butyl alcohol
by hydrolysing t-butyl chloride:

$$Me_3C-Cl \longrightarrow Me_3C^+ \quad ^-OH \longrightarrow Me_3C-OH$$

Draw the mechanism of the imaginary reverse
reaction, the formation of t-butyl chloride
from the alcohol.

---

2. $Me_3C-OH \Longrightarrow Me_3C^+ \quad ^-Cl \Longrightarrow Me_3C-Cl$

This then is the disconnection corresponding
to the reaction. It is the thinking device
we use to help us work out a synthesis of
t-butyl alcohol. We could of course have
broken any other bond in the target molecule
such as:

$$\begin{matrix} Me \\ Me-C-OH \\ Me \end{matrix} \Longrightarrow \begin{matrix} Me^+ \\ \quad \\ Me \end{matrix} \quad \begin{matrix} \\ ^-C-OH \\ Me \end{matrix}$$

Why is this less satisfactory than the disconnection at the start of this frame?

---

3. Because the intermediates Me^+ and Me_2COH^-
are pretty unlikely species and they would
have to be intermediates in the real reaction
too! We have already found the first way to
recognise a good disconnection: it has a
reasonable mechanism. Choose a disconnection
for this molecule, target molecule 3 (TM3)
breaking bond <u>a</u> or <u>b</u>. Draw the arrow and

the intermediates.

$$\text{TM3} \qquad \text{Ph} \underset{a}{\overset{}{\cancel{}}} \text{CH}_2 \underset{b}{\overset{}{\cancel{}}} \text{CH(CO}_2\text{Et)}_2$$

4.   The best one is <u>b</u>:

$$\text{PhCH}_2 \overset{\curvearrowleft}{\frown} \text{CH(CO}_2\text{Et)}_2 \Longrightarrow \text{PhCH}_2^+ + {}^-\text{CH(CO}_2\text{Et)}_2$$

since it gives a good cation and a good anion.
You have probably noticed the sign ($\Longrightarrow$) we
use for disconnections. This reminds us that
we are drawing the reverse of the real
reaction. Our synthesis of TM3 is then a
normal malonate reaction:

$$\text{CH}_2(\text{CO}_2\text{Et})_2 \xrightarrow{\text{EtO}^-} (\text{EtO}_2\text{C})_2\text{CH}^- \overset{\curvearrowright}{\frown} \overset{\overset{\text{Ph}}{\mid}}{\text{CH}_2} \overset{}{\frown} \text{Br} \rightarrow \text{TM3}$$

5.   Another class of reaction where you can
see at once that the disconnection is the
reverse of the reaction is Pericyclic
Reactions. An example would be the Diels-
Alder reaction between butadiene and maleic
anhydride. Draw the mechanism and the
product.

6.

TM6

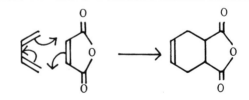

Now draw the disconnection (with mechanism)
on the product, TM6.

7.

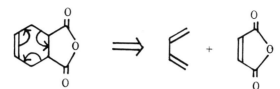

The double bond in the six-membered ring
showed us where to start the disconnection.
Can you see a similar disconnection for TM7?

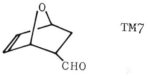

TM7

8.

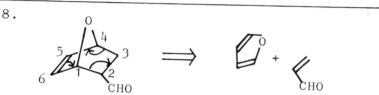

All you had to do was to find the six-membered
ring (numbered) containing the double bond and
draw the arrows.

---

9. So we shall be using disconnections corresponding to ionic and pericyclic reactions,
and we shall be looking all the time for a
good mechanism to guide us. You should now
see what a disconnection means and be ready
for the next stage. In the next few chapters
we shall study some important one group disconnections - reliable disconnections we can
use almost any time we see one particular
functional group in a target molecule.

B. ONE GROUP DISCONNECTIONS   13

1. DISCONNECTIONS OF SIMPLE ALCOHOLS

10. Simply by looking for a good mechanism,
you should be able to suggest a good disconnection for this alcohol:

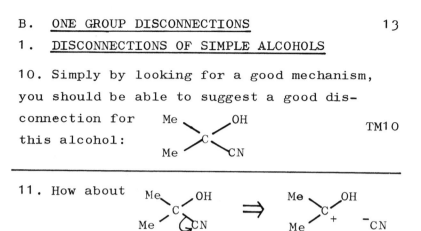

TM10

---

11. How about

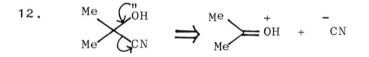

Cyanide is a good anion, and the cation is
stabilised by a lone pair of electrons on
oxygen. Draw the disconnection again using
the lone pair.

12.

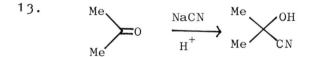

What is the real reaction which is the reverse
of this disconnection?

13.

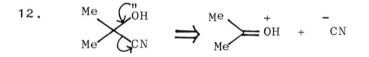

You probably saw the reaction before you saw
the disconnection! All simple alcohols can
be disconnected in this way. We simply
choose the most stable anion of the substituents and disconnect to a carbonyl compound:

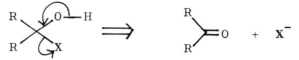

Suggest a disconnection for TM13:

TM13

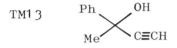

---

14. The acetylene anion $HC \equiv C^-$ is the most stable so:

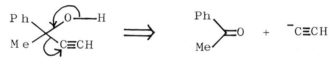

What is the real reaction?

15.

$$CH \equiv CH \xrightarrow[\text{liquid } NH_3]{\text{Na}} CH \equiv C^- \xrightarrow{\text{PhCOMe}} TM13$$

More usually, none of the substituents gives a stable anion and so we use the synthetic equivalent of the anion - the Grignard reagent or alkyl lithium. - We refer to "Et^-" as a SYNTHON for which EtMgBr is the synthetic equivalent.

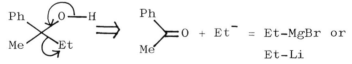

Draw the real reaction, the reverse of this disconnection, using EtLi with the mechanism.

16.

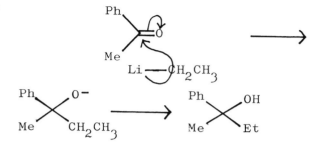

You can see how the alkyl-lithium acts as the
synthon $CH_3CH_2^-$ since the carbon-lithium bond
breaks so that the electrons go with the
carbon atom. Suggest
a disconnection for   TM16
TM16.

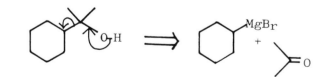

---

17. There are two possibilities:

a)

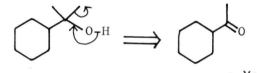

+ MeMgI

b)

Both have reasonable mechanisms, but we prefer (b) because it introduces more simplification. Route (a) simply chops off one
carbon atom and leaves us with a new target
almost as difficult to make as TM16. Route
(b) however breaks the molecule into two more

equal pieces – acetone and cyclohexyl bromide.

   We now have two criteria for a good disconnection: we look for (a) a good mechanism and (b) the greatest simplification.

---

18. An alternative approach to this problem, providing two of the groups on the tertiary alcohol are the same, is to remove both in a single disconnection going back to an ester and two mols of the Grignard reagent:

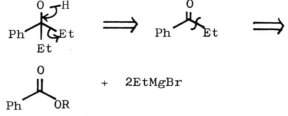

and the reaction is

$$PhCO_2R \xrightarrow{\quad 2\ EtMgBr \quad} PhC(Et)_2OH$$

How would you make TM18?

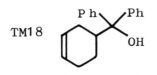

TM18

19.

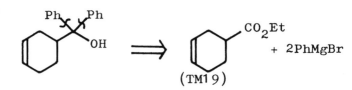

Can you continue one stage further back from TM19?

20. TM19 has a double bond in a six-membered ring and we can use the Diels-Alder disconnection (frames 5-8).

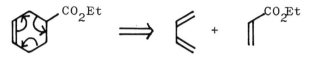

Note that the Diels-Alder reaction works best when there is an electron-withdrawing group (here CO_2Et) on the olefinic component.

---

21. If one of the groups in the alcohol carbon atom is H, then another disconnection is:

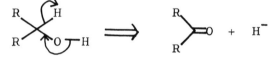

The synthetic equivalents of the synthon H^- are the hydride donors sodium borohydride $NaBH_4$, and lithium aluminium hydride $LiAlH_4$. How might you make TM21 using this disconnection?

---

22. Remove either one or both hydrogen atoms:

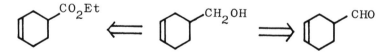

either starting material can again be made by a Diels-Alder reaction.

The complete syntheses are then:

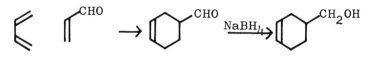

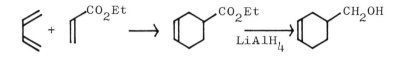

Note that $NaBH_4$ reduces aldehydes (and ketones) but not esters while $LiAlH_4$ reduces just about all carbonyl compounds. Neither reagent reduces an isolated double bond.

2.   COMPOUNDS DERIVED FROM ALCOHOLS

23. Have you noticed that the disconnections involving H^- are simply redox reactions and do not alter the carbon skeleton of the molecule? They are not then really disconnections at all but <u>Functional Group Interconversions</u> or FGI for short.

   Alcohols are key functional groups in synthesis because their synthesis can be planned by an important disconnection and because they can be converted into a whole family of other functional groups. List three types of molecule you might make from an alcohol by FGI.

24. You might have chosen any from this chart: (there are others)

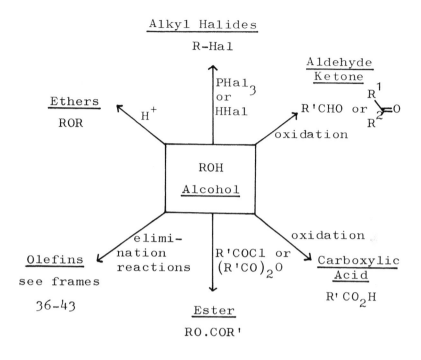

25. These FGI's are mostly straightforward, and the synthesis of any of these compounds is often best analysed by first going back to the alcohol and then disconnecting that. How would you make TM25?

TM25

26. <u>Analysis</u>:

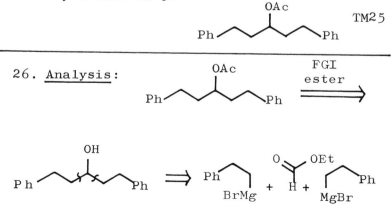

Synthesis:

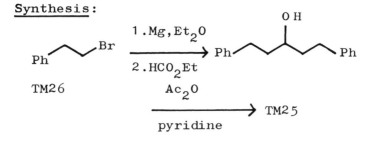

TM26

Ac$_2$O
——————————→ TM25
pyridine

---

27. But let us analyse the synthesis of the
halide (TM26) a bit more. The obvious way
to make it is:

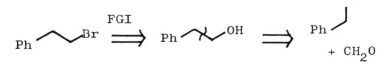

Unfortunately this route gives only a 40%
yield (<u>J. Amer. Chem. Soc.</u>, 1951, <u>73</u>, 3237)
in the Grignard reaction, largely because
benzyl Grignard reagents easily give radicals
which polymerise. In any case, it's poor
tactics to chop off carbon atoms one at a
time, and a better disconnection would be:

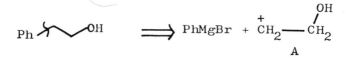

The reagent for synthon A is an epoxide so
that the reaction becomes:

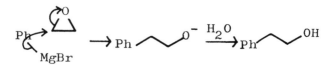

This reaction works well with monosubstituted epoxides:

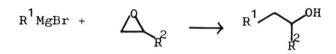

but is unreliable if there are more sub-stituents as you will see.

---

3.   REVIEW PROBLEMS

28. From time to time during the programme, I shall break off from introducing new ideas and help you consolidate what you've already learnt with some review problems.   These are meant to be realistic problems showing why synthesis is important and should let you try out your growing skills.   You can either do the review problems as you meet them or come back later and use them as revision material or combine both methods by doing one or two now and the rest later.   These remarks apply to all the review problems and I won't repeat them each time.

---

29. Review Problem 1: In 1936, Robinson carried out this reaction, hoping to get the alcohol A:

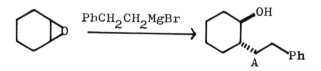

He got an alcohol all right, but it clearly

TM29

OH

Ph

wasn't A, and he
thought it might be
TM29.

He therefore wanted to synthesise TM29 to
check.   Even with modern spectroscopic methods
the quickest way to check the identity of a
compound will often be to synthesise it by an
unambiguous route and compare the n.m.r. and
fingerprint i.r. spectra.   How then would you
make TM29?

30. <u>Analysis</u>: the obvious disconnection takes
us back to the halide used by Robinson,
the one we synthesised in frame 27:

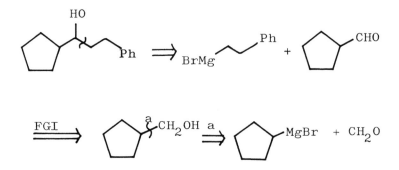

This time the one-carbon disconnection <u>a</u> is
all right because the Grignard reagent is
from a normal alkyl halide and does not
polymerise.

Synthesis:

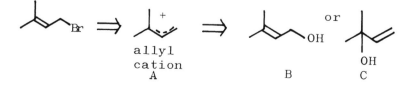

TM29, made by this route, did indeed turn out
to be identical with the compound Robinson
had made, and you might like to work out how
it was formed.   The reaction is discussed in
Norman, p.501.   The synthesis is described in
J. Amer. Chem. Soc., 1926, 48, 1080; Tetrahedron Letters, 1975, 2647, and Robinson's
original paper J. Chem. Soc., 1936, 80.

---

31. Review Problem 2:

This allyl bromide is   TM31
an important inter-
mediate in the synthesis of terpenes (includ-
ing many flavouring and perfumery compounds),
as the five carbon fragment occurs widely in
nature.   How would you make it?

---

32. Analysis: did you consider both possible
allylic alcohols as precursors?

Both give TM31 on treatment with HBr as the
cation A reacts preferentially with Br⁻ at the
<u>less</u> substituted carbon atom to give the <u>more</u>
substituted double bond.   Think again.

---

33. <u>Analysis</u>: you could make 32B by using a
vinyl Grignard reagent and formaldehyde but it
is easier to go via 32C and use the acetylide
ion (frames 14-15) as a reagent for the
synthon ⁻CH=CH₂:

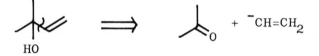

<u>Synthesis</u>: partial reduction of the
acetylene gives the olefin:

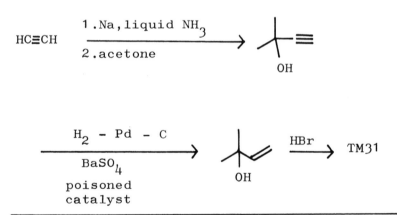

---

34. <u>Review Problem 3</u>: This odd-looking molecule (TM34) was used by Corey as an intermediate in the synthesis
of maytansine, an anti-
tumour compound.

TM34

How would you make it? Don't be deceived by
its oddness - identify the functional group
and you will see what to do first.

---

35. <u>Analysis</u>: the functional group is an
acetal derived from alcohols and a carbonyl
compound.

    The diol must have a cis double bond so
we can use the acetylene trick again here.

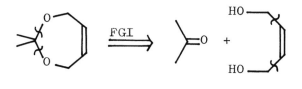

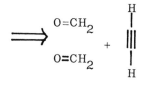

<u>Synthesis</u> (<u>Tetrahedron Letters</u>, 1975,
2643):

H —≡— H $\xrightarrow[\text{2.base,CH}_2\text{O}]{\text{1.base,CH}_2\text{O}}$

HO ⟍ ||| ⟋ HO $\xrightarrow[\text{BaSO}_4]{\text{H}_2\text{-Pd-C}}$

HO ⟍ ⟍ ⟋ HO $\xrightarrow[\text{H}^+]{}$ TM34

4. DISCONNECTIONS OF SIMPLE OLEFINS

36. Olefins are a little more complicated to analyse than alcohols. They can be made by the dehydration of alcohols:

$$Me_3C\text{-}OH \xrightarrow{\;H^+\;} Me_2C{=}CH_2 \;+\; H_2O$$

So the FGI stage in designing an olefin synthesis is to add water across the double TM36 ⬡Ph   bond. How would you synthesise TM36?

---

37. You should have two possible alcohols as the next step back, choosing one of these because it gives a useful disconnection while the other does not.

---

38. <u>Analysis:</u>

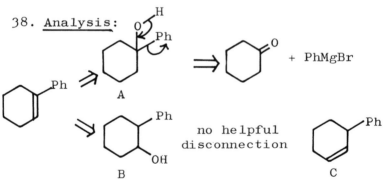

We must also consider whether the dehydration reaction might be ambiguous. Thus A can give only TM36 on dehydration but B might TM38 ⟍⟋—Ph   give C as well. How would you make TM38?

---

39. The alternatives are:

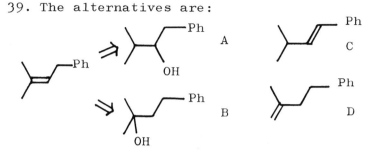

Dehydration of A could also give C, the conjugated olefin, but dehydration of B will give
only TM38 and none of the less substituted D.
Now finish off the analysis and write out the
synthesis.

---

40. <u>Analysis</u>:

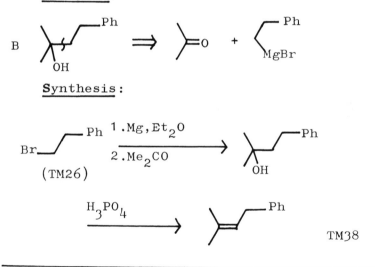

41. An alternative route to olefins is by an
immediate disconnection of the double bond.
This corresponds to the <u>Wittig reaction</u>:

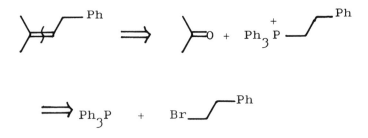

If you are unfamiliar with the Wittig reaction, see Norman p.297-299 or Tedder, Part 3, p.233-6.

The advantages of this route are that it is very short and that the double bond must go where we want it. Otherwise it is very like the route in frame 40 and actually uses the same starting materials. How might you make TM41?

TM41

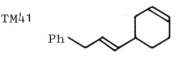

42. Choosing to disconnect the double bond outside the ring, as this will give us two fragments:

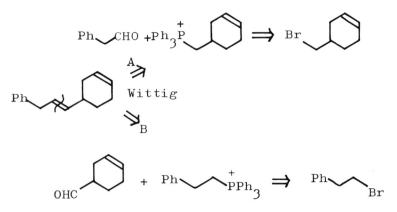

The starting materials for route B are
recognisable as the halide we used in frame
41 and an aldehyde easily made by a Diels-
Alder reaction.   The other route could also
be used but the starting materials are not so
readily available.   Write out the complete
synthesis.

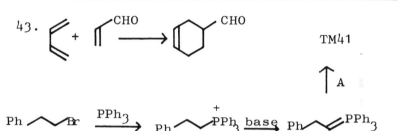

43.

TM41

↑ A

This is a good opportunity to mention our
third criterion for a good disconnection -
that it leads to recognisable starting
materials.   We have used this criterion
already in frames 20 and 42.

5.   DISCONNECTIONS OF ARYL KETONES

44. The Wittig reaction is important enough
to be our second major one group disconnection.   The first was the disconnection of
alcohols to carbonyl compounds and Grignard
reagents.   Our third major one disconnects
the bond joining an aromatic ring to an
aliphatic side chain.   So we would make TM44
by the Friedel-Crafts reaction using acetyl
chloride and aluminium chloride to attack the
benzene ring:

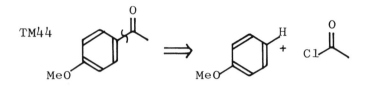

TM44

---

45. In principle we can disconnect any bond
next to an aromatic ring in this way, though
not always in practice. How would you make

TM45?

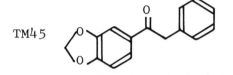

TM45

---

46. One of the two possible disconnections <u>a</u>
is better as it gives us an acyl rather than
an alkyl halide and an activated benzene ring.

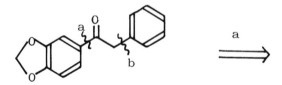

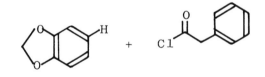

If you're not sure about why this is so, or
don't understand the mechanism of the Friedel-
Crafts reaction, you will find help in Tedder,
part 2, pages 212-215 or Norman, pages 363-
370.

47. Sometimes a choice between two disconnections of this sort can be made by our first
criterion (a good mechanism). How would you
make TM47?

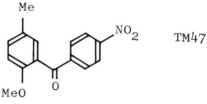

TM47

---

48. There are two possible disconnections:

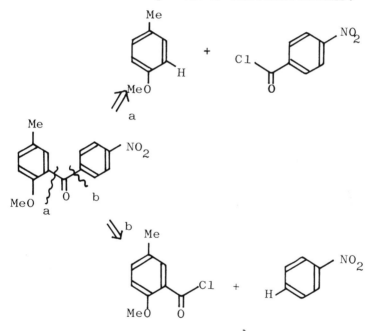

Disconnection **b** will not do as the nitro group
is meta-directing and in any case nitro benzene will not react under Friedel-Crafts conditions. Disconnection **a** is fine as the MeO
group is more powerfully ortho-directing than
the Me group (Ber., 1907, 40, 3514).

6.   CONTROL

49. Before we complete the disconnections of
carbonyl compounds we shall look at some
aspects of control in synthesis as a break
from the systematic analysis.

   Why might the obvious disconnection on
TM49 give trouble when the real reaction is
tried?

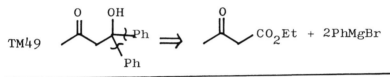

50. The Grignard reagent might first attack
the ketone giving the wrong product.

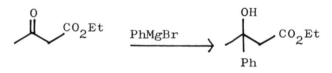

   To stop this we <u>protect</u> the ketone by a
reversible FGI. A common method is to make
the cyclic ketal:

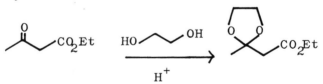

   Now complete the synthesis.
If you're not sure of the mechanism of acetal
formation or just want to know a bit more
about acetals, read frames 1-21 and 62-64 of
the Carbonyl Programme.

51.

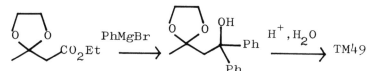

Any functional group can act as a protecting group providing it can easily be added
and removed and providing of course that it
doesn't react with the reagent! We shall meet
more examples as we work through the programme.

---

52. Sometimes, rather than <u>protect</u> one part
of a molecule, it is better to <u>activate</u>
another.

This reaction gives only a poor yield. Why?
Enolisation is involved: if you're not sure
about this, see frames 169 ff of the Carbonyl
Programme.

---

53. Because the product is at least as reactive as the starting material, and further
reaction occurs:

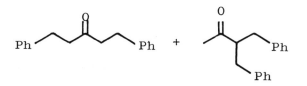

We can't protect the carbonyl group without
stopping the reaction, so we <u>activate</u> one
position by adding a CO_2Et group and using the
ester A below, the synthetic equivalent of
acetone, instead of acetone itself.   Here is
the reaction; draw a mechanism for it.

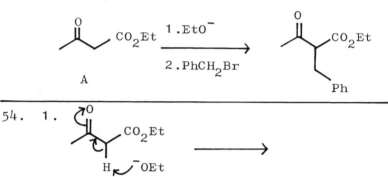

54.   1.

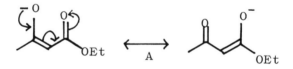

only this stable enolate A formed

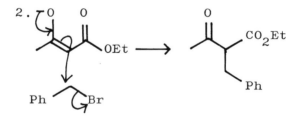

Of course, we must now remove the activating
group, CO_2Et in this case, just as we had to
remove the protecting group before.   How might
we do this?

55. By hydrolysis and decarboxylation:

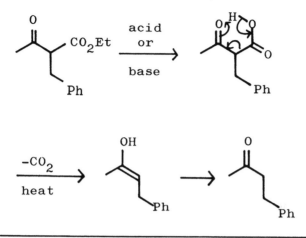

56. This is then a general synthesis for
ketones and the corresponding disconnection is

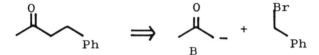

The acetoacetate enolate ion (A in frame 54)
is a reagent for the synthon B, the acetone
anion.   We shall discover how to add the CO_2Et
activating group later.

57. Protection and activation give us a
reagent for the synthon $^-CH_2CO_2H$.   We protect
the acid as an ester and add another ester
group as activation, giving malonic ester:
$CH_2(CO_2Et)_2$.   How
would you make TM57?

TM57

58. <u>Analysis</u>:

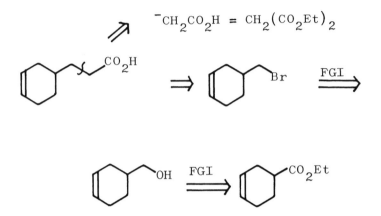

$$^-CH_2CO_2H = CH_2(CO_2Et)_2$$

Choosing this disconnection because we
recognise a starting material easily made
by a Diels-Alder reaction (cf. frame 22).

<u>Synthesis</u>:

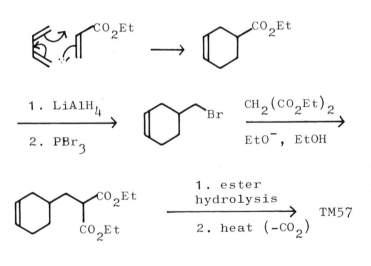

59. Here is quite a difficult problem: to
solve it you will need to use both protection
and activation. Two hints: the disconnections
are shown and you might like to start by thinking how you would make a <u>cis</u> olefin. How can
you make TM59?

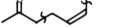

TM59

---

60. <u>Analysis</u>: This <u>cis</u> olefin will presumably
come from an acetylene: we can then use an
acetylide anion:

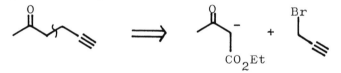

+ MeI

Now we can disconnect the ketone using our
synthetic equivalent for the acetone anion:

<u>Synthesis</u>: (Crombie, <u>J. Chem. Soc.</u> (C), 1969,
1016). The acetylenic bromide corresponding
to allyl bromide is called propargyl bromide
and is reactive and readily available. We
shall need to protect the ketone before we
make the acetylene anion. It turns out that
protection and decarboxylation can be done in
one step.

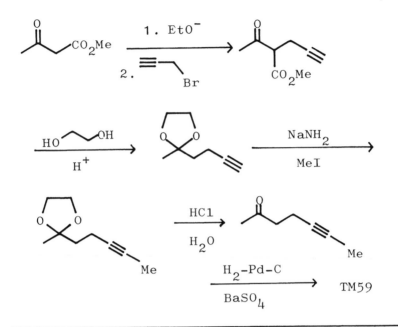

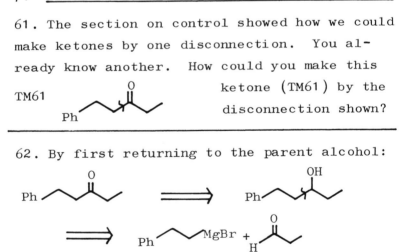

7.   DISCONNECTIONS OF SIMPLE KETONES AND ACIDS

61. The section on control showed how we could
make ketones by one disconnection.   You already know another.   How could you make this

TM61   ketone (TM61) by the

                              disconnection shown?

62. By first returning to the parent alcohol:

63. You also know how to make acids by FGI
from a primary alcohol; but an acid is itself

a hydroxyl compound and can be disconnected in the same way as alcohols. What do you get if you do this:

$$R \rightarrow \overset{O}{\underset{}{\overset{\|}{C}}} - OH$$

---

64. $R - \overset{O}{\underset{}{\overset{\|}{C}}} - O - H \implies R - MgBr + CO_2$

Acid derivatives are made directly from acids or by conversion from other acid derivatives depending on their stability. The most important are esters (RCO_2Et), amides (RCO_2NR_2), anhydrides $(RCO.O.COR)$ and acid chlorides $(RCOCl)$. Arrange these in an order of stability, the most reactive at the top of the list, the most stable at the bottom.

---

65. RCOCl   most reactive
     RCO.O.COR
     RCO.OR'
     RCONR$'_2$ most stable

Conversions down the list are easy - simply use the appropriate nucleophile. Thus:

$$RCOCl \xrightarrow{\text{R'OH}} RCO.OR' \xrightarrow{\text{R}_2\text{NH}} RCO.NR'_2$$

All can be hydrolysed to the acid, and the list can be entered at the top from the acid by using $SOCl_2$ or PCl_5 to make the acid chloride.

---

66. This gives us a complete chart for acid derivatives.

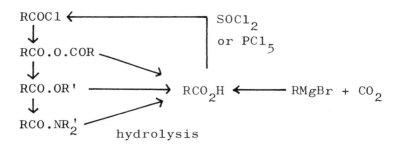

---

67. So how could you make this acid

                    derivative?

TM67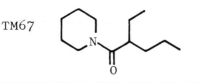

---

68. <u>Analysis</u>: Amide ∴ FGI back to carbo-xylic acid:

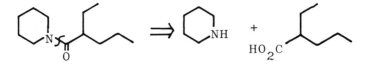

acid is branched at α-carbon ∴ use discon-nection of α bond

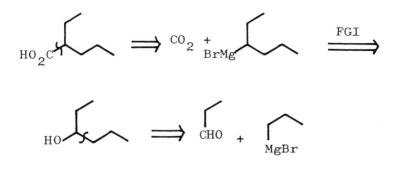

Synthesis:

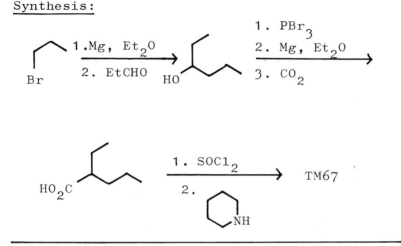

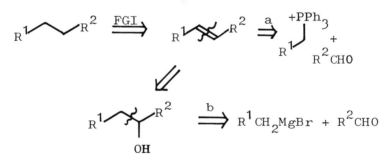

69. Finally in our treatment of one group disconnections we ought to consider how to synthesise fully saturated hydrocarbons - compounds
with no FG at all! These are often made by
hydrogenation of a double bond, and so the
disconnection can be made anywhere we like:

Using our principles for good disconnections,
we shall obviously look for two roughly equal
recognisable fragments. So how would you make
TM69?

70. <u>Analysis</u>: There are many answers.   One is
to put the double bond as close to the benzene
ring as possible:

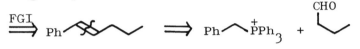

   <u>Synthesis</u>:

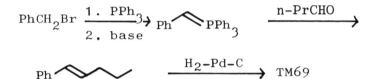

 $\xrightarrow{\text{H}_2\text{-Pd-C}}$ TM69

71. A guide we can sometimes use, particularly
if we use disconnection <u>b</u> in frame 69, is to
put the OH group at a branch point in the
molecule, knowing that disconnection will be
easy there.   Try this:

TM71

72. <u>Analysis</u>: Use branch point ● as a guide.

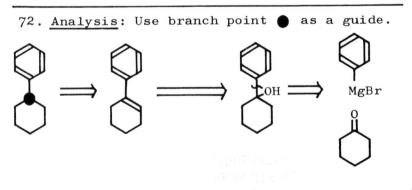

Synthesis:

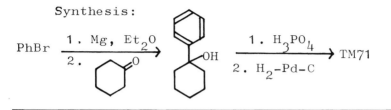

PhBr $\xrightarrow[\text{2.}]{\text{1. Mg, Et}_2\text{O}}$ [cyclohexanone] $\xrightarrow[\text{2. H}_2\text{-Pd-C}]{\text{1. H}_3\text{PO}_4}$ TM71

---

8.   SUMMARY AND REVISION

73. Although you have analysed the synthesis
of many compounds and considered mechanisms
of many reactions, we have collected only a
handful of important one group disconnections.
Can you fill in the details of these:

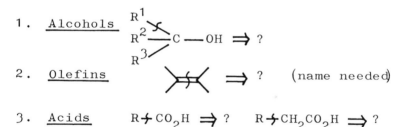

1.   <u>Alcohols</u> $R^2 \!-\! C \!-\! OH \Rightarrow$?   with R^1, R^3

2.   <u>Olefins</u> $\Rightarrow$?   (name needed)

3.   <u>Acids</u> $R \!+\! CO_2H \Rightarrow$? $R \!+\! CH_2CO_2H \Rightarrow$?

4.   Carbonyl Compounds $Ar \!+\! COR \Rightarrow$ (name needed)

$R \!+\! CH_2COR$

74.

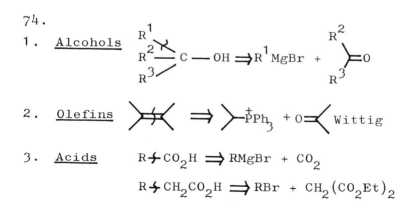

1.   <u>Alcohols</u> $R^2 \!-\! C \!-\! OH \Rightarrow R^1 MgBr +$ R^2, R^3 >=O   with R^1, R^3

2.   <u>Olefins</u> $\Rightarrow$ $\overset{+}{P}Ph_3$ + $O =$ Wittig

3.   <u>Acids</u> $R \!+\! CO_2H \Rightarrow RMgBr + CO_2$

$R \!+\! CH_2CO_2H \Rightarrow RBr + CH_2(CO_2Et)_2$

4. Carbonyl Compounds

$$Ar + COR \implies ArH + ClCOR \quad \text{Friedel-Crafts}$$

$$R + CH_2COR \implies RBr + EtO_2CCH_2\overset{O}{\overset{\|}{C}}R$$

---

75. We have three or four ways to recognise a good disconnection. Make a list of these.

76.

1. Good mechanism.

2. Greatest possible simplification.

3. Gives recognisable starting materials.

(You might also have mentioned the use of the branch point as a guide).

---

77. A useful thing to do at this stage would be for you to start making a chart, of your own design and for your own use, showing all the useful synthetic links between the various classes of compound. You can then add to this later but a colourful, well designed chart of the relationships between single functional groups is a good reference.

---

9. REVIEW PROBLEMS

78. Review Problem 4: This compound (TM78) is an important intermediate in the synthesis of alkaloids: Treatment with $POCl_3$ gives the poppy alkaloid papaverine. How would you make TM78 from simple starting materials?

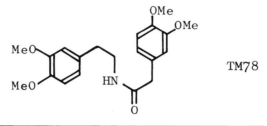

TM78

79. <u>Analysis</u>: There are four ether groups,
but they are peripheral and easily made.
The key FG is the amide which we must disconnect first at the C-N bond. Both acid
and amine could be made from the same nitrile.

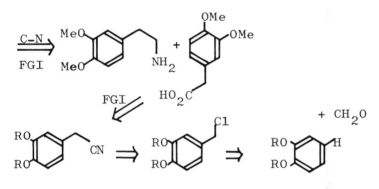

<u>Synthesis</u>: from readily available
catechol:

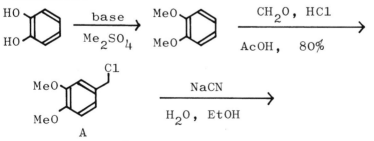

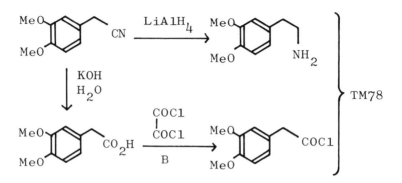

The reaction giving A is 'chloromethylation',
a reliable method of adding a CH_2OH equivalent
to an aromatic ring. You may have been surprised at the use of reagent B to make an
acid chloride. B is oxalyl chloride and is
often used when pure acid chlorides are wanted
- the other products are gases (which?).

The nitrile is described in a patent
(Chem. Abs., 1955, 15963); the last stages
were carried out by A. R. Battersby's research
group at Cambridge. Chloromethylation is
described in Tedder, Vol 2, p.213 and Norman
p.372-3.

---

80. Review Problem 5: 'Brufen' (TM80), Boots
anti-rheumatic compound, is one of Britain's

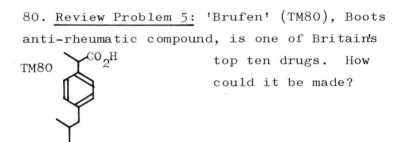

top ten drugs. How
could it be made?

---

81. Analysis: The carboxylic acid is the only
FG so we can start there:

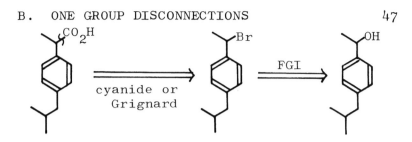

We now have a benzyl alcohol so we use
Friedel-Crafts rather than Grignard:

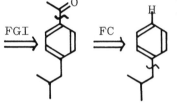

Again we want to use
Friedel-Crafts but we
must use acylation
rather than alkylation
or we shall get rearr-
angement.

Synthesis: This is one possible approach - we
don't actually know how it is done.

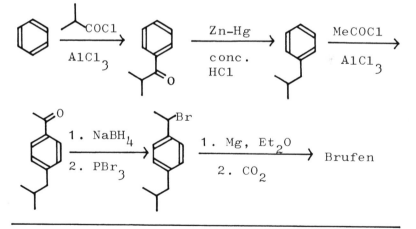

82. Review Problem 6: Some chemists who were
investigating the possibility of reversible
Friedel-Crafts reactions, wanted an activated
aromatic ring connected to a branched alkyl

chain and chose to make TM82.   How would you
do it?

TM82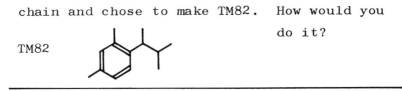

---

83. <u>Analysis</u>: Using the branch-point in the
largest side chain as a guide, we can put in
a hydroxyl group (as in frame 72).

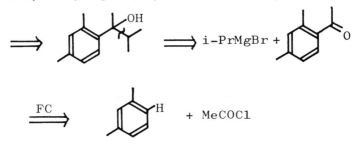

$\Rightarrow$ i-PrMgBr +

$\xrightarrow{\text{FC}}$ + MeCOCl

   <u>Synthesis</u>: note that there is only one
activated site for the Friedel-Crafts reaction
- <u>o</u> and <u>p</u> to the two methyl groups but not
between them for steric reasons:

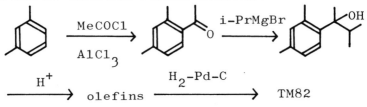

$\xrightarrow{\text{MeCOCl}}{\text{AlCl}_3}$

$\xrightarrow{\text{H}^+}$ olefins $\xrightarrow{\text{H}_2\text{-Pd-C}}$ TM82

This was essentially the method used by the
chemists who went on to investigate the
chemistry of TM82 (<u>J. Org. Chem.</u>, 1942, <u>7</u>, 6).

1. 1,3-DIOXYGENATED SKELETONS
(a) β-HYDROXY CARBONYL COMPOUNDS

84. When a molecule contains two functional
groups, the best disconnection uses the two
together. So if you consider TM84 as an
alcohol, and use the carbonyl group to guide
your disconnection,
what do you get?

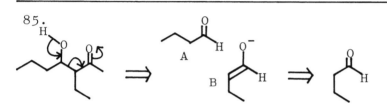

TM84

85.

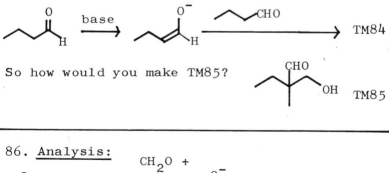

The anion B is just the enolate anion of a
carbonyl compound, actually the same as A.
So there is no need to use a Grignard reagent
or any other synthetic equivalent in this
reaction: anion B itself can be the intermediate and we simply treat the aldehyde with
mild base:

So how would you make TM85?

TM85

86. <u>Analysis:</u>

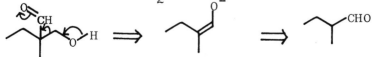

Synthesis:

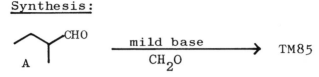

A

mild base
─────────────→   TM85
CH₂O

You may wonder why aldehyde A doesn't react
with itself but reacts instead with formaldehyde.   This is just one aspect of control
in carbonyl condensations, treated thoroughly
in frames 217-315 of the Carbonyl Programme.
In this case, only aldehyde A can enolise but
formaldehyde is more electrophilic.

TM86

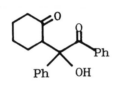

Now try this problem:
How would you
synthesise TM86?

───

87. Analysis: It may be tempting to disconnect bond **a** but this would give the un-
known and presumably very unstable PhC=O⁻
synthon.   The better disconnection is bond **b**
giving two carbonyl compounds.

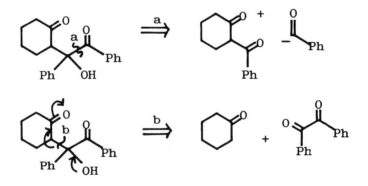

Synthesis: Only cyclohexanone can enolise, but the α-diketone is more electrophilic - no control needed:

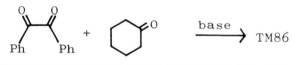

(b) α,β-UNSATURATED CARBONYL COMPOUNDS

88. Using one of our methods of analysing the synthesis of olefins, that is FGI to an alcohol, write down both the alcohols from which you might make TM88 and see which you prefer.

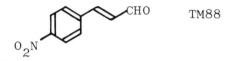

89. FGI (b) gives an alcohol we can disconnect easily:

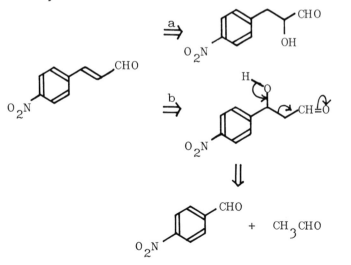

Synthesis: The synthesis uses rather more
vigorous conditions than those which gave the
β-hydroxy carbonyl compounds. In fact (Bull.
Chem. Soc. Japan, 1952, 25, 54, Chem. Abs.,
1954, 48, 5143) you can either treat the β-
hydroxy compound with HCl in acetic acid or do
the condensation in base:

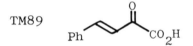 excess   MeCHO
$\xrightarrow{\hspace{3cm}}$ TM88
KOH, MeOH

Only the acetaldehyde can enolise but the two
aldehydes are about equally electrophilic: we
use an excess of acetaldehyde to compensate
for its self-condensation. What about TM89?

TM89

90. Analysis: by the same method:

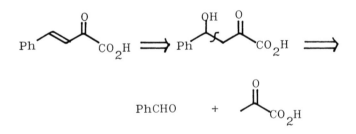

Synthesis: No control is need because only
the ketoacid can enolise and the aldehyde is
more electrophilic. TM89 is formed in 80%
yield when the two starting materials are

mixed in MeOH with KOH at room temperature
(<u>Helv. Chim. Acta</u>, 1931, <u>14</u>, 783).

---

91. So we can disconnect <u>any</u> α,β-unsaturated
carbonyl compound along the double bond,
writing CH_2 at one end and C=O at the other.

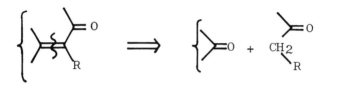

How about this one:

TM91

---

92. I hope you weren't put off by the ring:

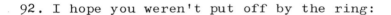

We shall discover how to synthesise this
starting material later.

---

93. So to summarise these two-group disconnections: we can always disconnect the α,β
bond in either of these structures:

Mild conditions (usually base) give the alcohol, more vigorous conditions (acid or base) give the enone.

(c) 1,3-<u>DICARBONYL COMPOUNDS</u>

94. Disconnection of the same bond gives a good synthesis of 1,3-dicarbonyl compounds:

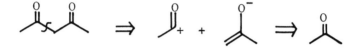

The reagent for the synthon RCO$^+$ will be RCOX where X is a leaving group, such as OEt. So how would you make TM94?

TM94

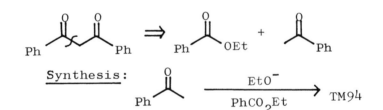

Ph   Ph

---

95. <u>Analysis</u>:

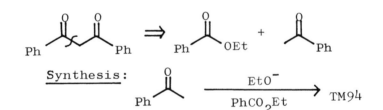

Ph   Ph

<u>Synthesis</u>:

Ph $\xrightarrow[\text{PhCO}_2\text{Et}]{\text{EtO}^-}$ TM94

Now what about TM95?

TM95

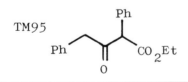

Ph CO_2Et

---

96. You had a choice between bonds <u>a</u> or <u>b</u>:

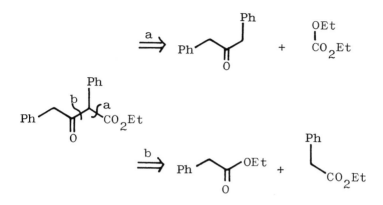

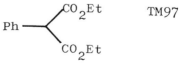

b has the advantage of greater simplification. It also has an advantage we used previously (in frame 85) that of symmetry: both starting materials are actually the same molecule. The synthesis is therefore the Claisen ester condensation.

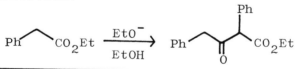

97. The other disconnection (a in frame 96) is very important if we want to add control in the form of a CO_2Et group. How would you make TM97?

Ph ⎯ $\overset{CO_2Et}{\underset{CO_2Et}{|}}$ TM97

98.   Simple:

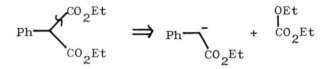

The compound $CO(OEt)_2$ is diethyl carbonate and is readily available. I hope you weren't seduced by the alternative:

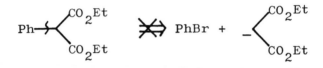

as nucleophilic substitutions on <u>aryl</u> halides need special conditions (see Tedder, vol 2, p.157 ff., or Norman, p.401 ff.). You may remember from frames 57-58 that TM97 is a reagent for $Ph\bar{C}HCO_2H$. How could we make TM98 using it?

TM98

99.   <u>Analysis</u>:

Synthesis:

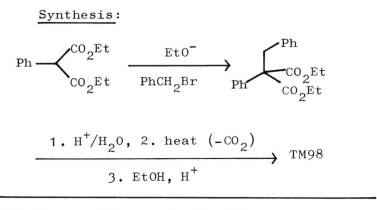

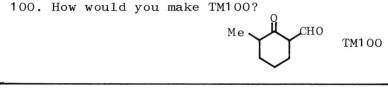

---

100. How would you make TM100?

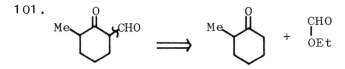

TM100

101.

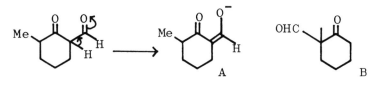

The one-carbon fragment is ethyl formate.
This reaction is important as a method of control since it occurs only on one side of the
carbonyl group: that is it is regioselective.
The reason is that this product can itself
enolise in

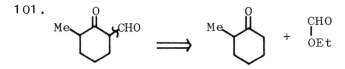

the basic reaction medium to form the stable
delocalised enolate A. This drives the
equilibrium over and would not be possible

with the alternative product B since there is
no hydrogen atom at the vital point.

How could we develop this into a synthesis of TM101?

TM101
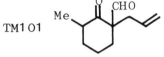

102.

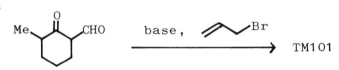

base, ⟋⟍⟋Br

⟶ TM101

Only the more stable enolate (101A) is formed
and this reacts well with allyl bromide. This
activating group (CHO) can be removed by base-
catalysed hydrolysis.   Mechanism?

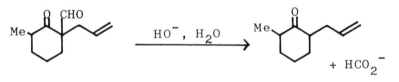

---

103. HO⁻ attacks the more reactive carbonyl
group:

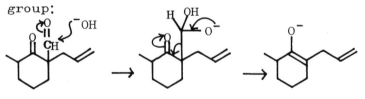

$+ HCO_2H$

More details on this are given in Section
V of the Carbonyl Programme, with a summary
in frame 324.

104. The one-carbon addition we used in frames
98 and 101 is all right if we just want to
add an activating group to a readily available
ketone, but is not otherwise good synthetic
practice!

    What alternative
disconnection is avail-
able here?

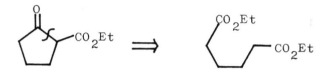

TM104

---

105. <u>Analysis</u>:

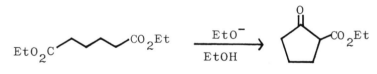

This starting material is symmetrical and
happens to be readily available. (See frame
194 ff).

    <u>Synthesis</u>: This cyclisation version of
the Claisen ester condensation is sometimes
called the Dieckmann Reaction.

$$EtO_2C \diagup\!\!\!\diagdown\!\!\!\diagup CO_2Et \xrightarrow[\text{EtOH}]{\text{EtO}^-} \text{(cyclopentanone)} CO_2Et$$

---

106. Sometimes we can be guided in our disconnection by the relative stabilities of the
possible anionic fragments.
How would you make TM106?

$$Ph \diagup\!\!\!\diagdown Ph$$
$$t\text{-}BuO_2C \qquad CO_2Bu\text{-}t$$

TM106

107. One disconnection gives a symmetrical
stable anion derived from an easily made malonate ester.

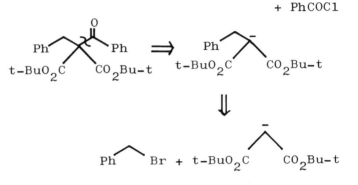

Synthesis:

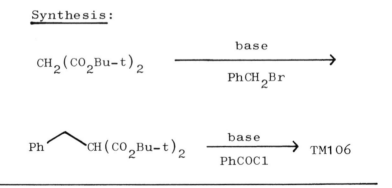

(d) <u>REVIEW PROBLEMS</u>

108. <u>Review Problem 7</u>: α-β-Unsaturated lactones are useful intermediates in synthesis as
they take part in Diels-Alder reactions to
build larger molecules
with more complex func-
tionality. How would
you make this one?

TM108

109. Analysis: We must first open the lactone ring:

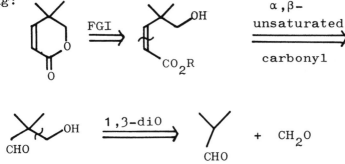

α,β-unsaturated carbonyl

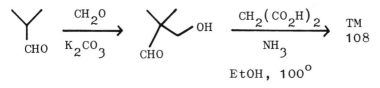

Synthesis: No control is needed in the first step: there is only one enolisable H atom on either aldehyde. If we use malonic acid for the second step, cyclisation and decarboxylation will be spontaneous (<u>Monatshefte</u>, 1904, <u>25</u>, 13).

$$\text{CHO} \xrightarrow[\text{K}_2\text{CO}_3]{\text{CH}_2\text{O}} \text{CHO} \xrightarrow[\substack{\text{NH}_3 \\ \text{EtOH, 100}^{\circ}}]{\text{CH}_2(\text{CO}_2\text{H})_2} \text{TM 108}$$

If you're not sure of the details of this last step, try working it out as a mechanistic problem.

___

110. <u>Review Problem 8</u>: Suggest a synthesis of the mydriatic (dilates the pupils of the eyes) cyclopentolate, TM110.

TM110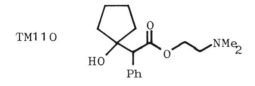

---

111. <u>Analysis</u>: We must first separate the ester into its component parts:

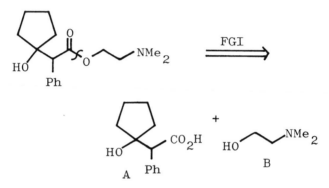

The alcohol B is a typical amine-epoxide adduct, and the acid A is a 1,3-dioxygenated compound:

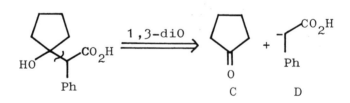

<u>Synthesis</u>: Control will be needed in the condensation as the ketone C is more reactive than the acid D both in enolisation and electrophilic power. The Reformatsky looks a good method. Again we don't know how this commercial product is actually made:

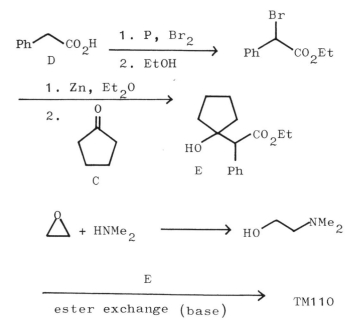

There is no danger that the tertiary alcohol group will form an ester under these conditions. Ester exchange is described in Norman p.134-5.

2. 1,5-DICARBONYL COMPOUNDS

112.So far in this section we have combined enolate anions with other carbonyl compounds by direct attack at the carbonyl group. We can expand the scope of this reaction by using α,β-unsaturated carbonyl compound electrophiles. This is the Mi Remind yourself of this by writ mechanism of a Michael reaction

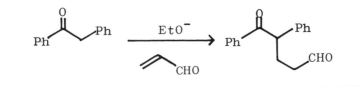

113.

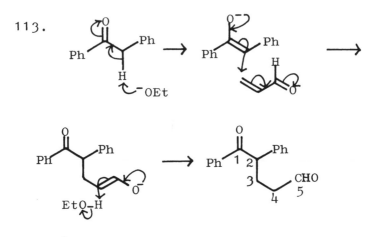

You see that this reaction makes a 1,5-di-
carbonyl compound: we can therefore disconnect
any such compound at either of the two middle
bonds.

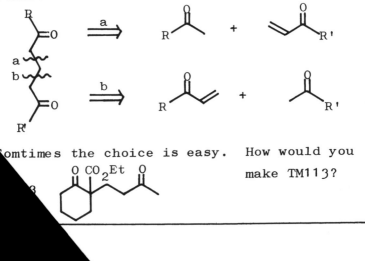

Somtimes the choice is easy. How would you
make TM113?

114. Only one of the two disconnections is possible:

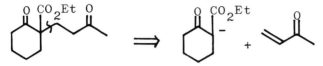

This disconnection is also good because:

(a)   it gives a stable anion

(b)   both starting materials can easily be made by methods outlined in frames 97-99 and 104-106.

Sometimes we must make a choice between two mechanistically reasonable disconnections. How about TM114?

TM114

115.

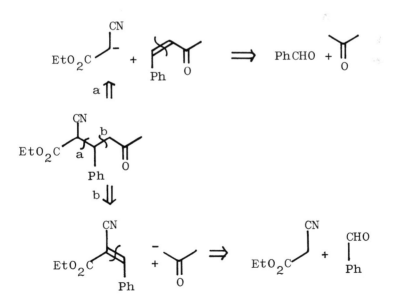

Both routes are acceptable and both get back
to the same three starting materials. Route <u>a</u>
uses a Michael reaction with a stable anion so
this is preferable.

---

116. The Michael reaction plays a part in some
more extended synthetic sequences of great
importance. Analyse TM116 as an α,β-unsatu-
rated carbonyl compound and continue your

analysis by the
Michael reaction.

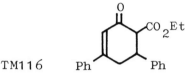

TM116

---

117. <u>Analysis</u>:

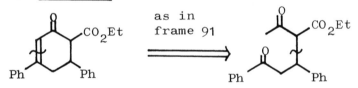

as in
frame 91

We now have a 1,5-dicarbonyl compound with
one good disconnection:

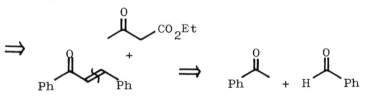

This sequence of Michael reaction and cyclisation is known as the Robinson annelation since it makes a ring.

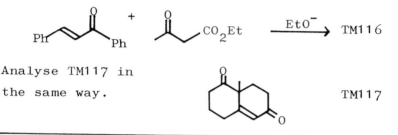

Analyse TM117 in the same way.   TM117

118. Starting as before with the α,β-unsaturated ketone:

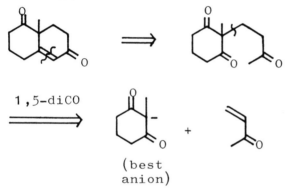

1,5-diCO

This sequence can be carried out in one or two steps and makes an important molecule for steroid syntheses. Further details are given in Fleming pages 59, 75 and 171 if you are interested.

119. If we want to make a simple 1,5-diketone we may have to use an activating group like CO_2Et to control the reaction. How would you make TM119?

TM119

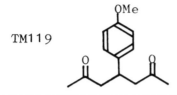

---

120. Analysis:

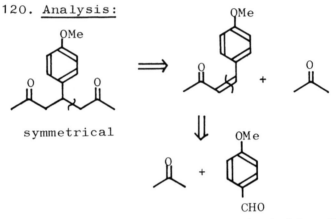

symmetrical

Synthesis: To ensure good yields, the reaction is best done on an activated compound, so the synthesis becomes:

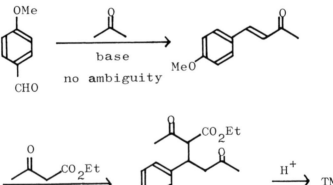

Now what about TM120?

TM120

---

121. <u>Analysis</u>:

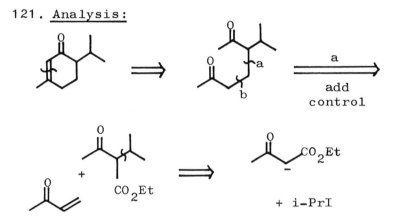

Choosing the Michael disconnection at <u>a</u> rather than <u>b</u> since we can then use the CO_2Et control group both for the alkylation and for the Michael reaction.

<u>Synthesis</u>:

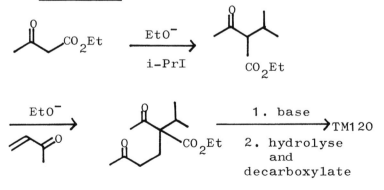

The final condensation could have gone the
other way too, but it doesn't, presumably
because attack on the other carbonyl group is
hindered.   TM120 is in fact piperitone, one
of the flavouring principles of mint, and has
been synthesised essentially by this route
(<u>J.C.S.</u>, 1935, 1585; <u>Rec. Trav. Chim.</u>, 1964,
<u>83</u>, 464; <u>Zhur. Obshchei Khim.</u>, 1964, <u>34</u>, 3092,
<u>Chem. Abs.</u>, 1964, <u>61</u>, 16098).

(a)   USE OF THE MANNICH REACTION

122. There is one special case worth discuss-
ing in some detail.   When vinyl ketones (e.g.
TM122) are needed for Michael reactions they
may obviously be made by the usual disconnection:

TM122

which gives formaldehyde as one of the starting materials.   Base-catalysed reactions with
this very reactive aldehyde often give poor
yields because of polymerisation and other
side reactions.   The Mannich reaction is used
instead:

Write a mechanism for this reaction.

123.

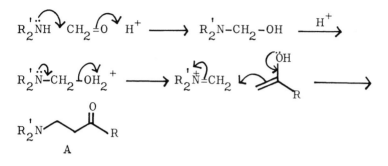

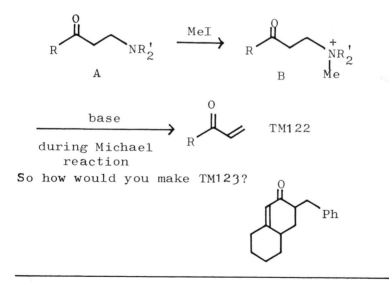

$R_2'N\!-\!CH_2\!-\!OH$

$R_2'N\!-\!CH_2\!-\!OH_2^+$

$R_2'\overset{+}{N}\!=\!CH_2$

A

Alkylation of the product (a 'Mannich Base' A)
gives a compound (B) which gives the required
vinyl ketone on elimination in base. This
last step is usually carried out in the basic
medium of the Michael reaction itself so that
the reactive vinyl ketone (TM122) need never
be isolated.

A   MeI   B

base
during Michael   TM122
reaction

So how would you make TM123?

124. <u>Analysis:</u>

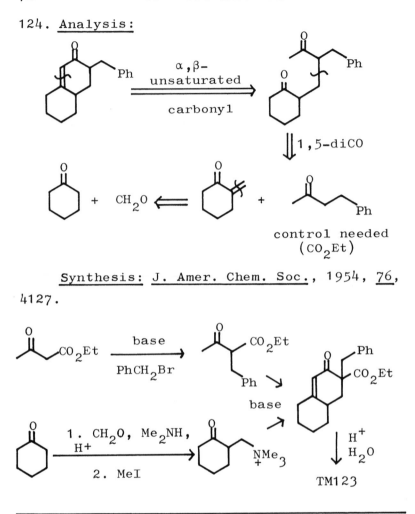

Synthesis: <u>J. Amer. Chem. Soc.</u>, 1954, <u>76</u>, 4127.

---

3.   <u>REVIEW PROBLEMS</u>

125. <u>Review Problem 9</u> - Suggest a synthesis of TM125, a commonly used synthetic intermediate called Hagemann's ester.

TM125

---

126. Analysis:

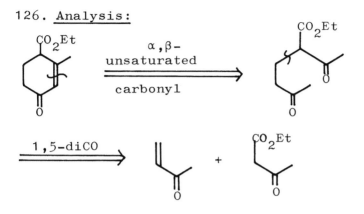

Synthesis: Though we could follow the stepwise pattern of the disconnections, it is easier to add an activating group to the acetone molecule so that our starting materials are two molecules of acetoacetate and formaldehyde. It turns out that Hagemann's ester can be made in two steps without having to alkylate the Mannich base:

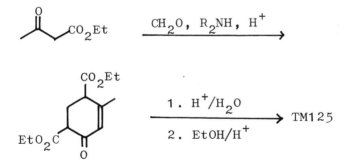

127. Review Problem 10 - Suggest a synthesis for

EtO$_2$C

TM127

128. <u>Analysis</u>: Disregarding the remote and unhelpful double bond, we can disconnect as a 1,3-dioxygenated compound (frames 94-107).

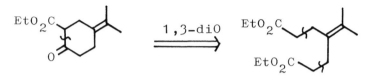

Now note symmetry. Doubly allylic disconnection keeps symmetry, requires activation (frames 57-8 and 101-2).

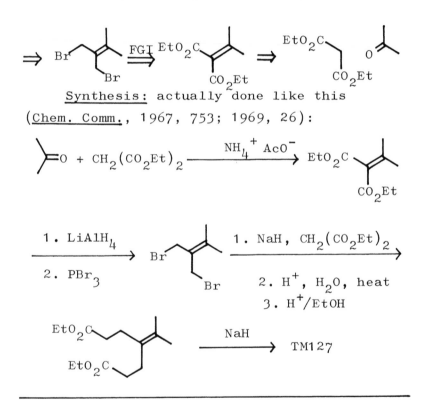

<u>Synthesis</u>: actually done like this (<u>Chem. Comm.</u>, 1967, 753; 1969, 26):

129. <u>Review Problem 11</u> - suggest a synthesis
for TM129

TM129

130. <u>Analysis</u>: Treat first as an α,β-unsatu-
rated ketone:

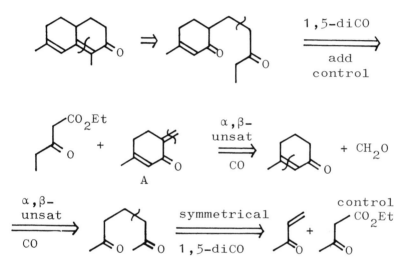

Synthesis: All by standard steps.
Though the Michael addition on A could in
theory occur at either double bond, the
unsubstituted position out of the ring is
much more reactive than the disubstituted
position in the ring and only the wanted
reaction occurs. <u>Bull. Soc. Chim. France</u>,
1955, 8.

1.   THE 1,2-DIOXYGENATION PATTERN
(a)   α-HYDROXY-CARBONYL COMPOUNDS

131. So far all our two group disconnections have sensible synthons with anions or cations all stabilised by functional groups in the right positions. This won't always be the case. Supposing we wanted to make the hydroxy-acid TM131; we could treat it as an alcohol:

but we get the apparently absurd synthon $^-CO_2H$. In fact, there is a common reagent for this synthon - a simple one-carbon anion which adds to ketones and whose adduct with A could easily be converted into TM131. What is it?

132. Cyanide ion! So the synthesis becomes:

133. The aldehyde or ketone needed for this reaction is not always readily available. TM133, labelled with radioactive ^{14}C in one carboxyl group, was needed for a biochemical labelling experiment. How would you make it?

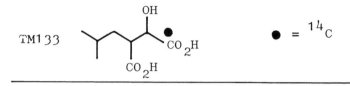

TM133

● = ^{14}C

134. <u>Analysis</u>: The α-hydroxy acid can best be
made from an aldehyde and $^{14}CN^-$, then we can
carry on as usual with a 1,3-dicarbonyl disconnection:

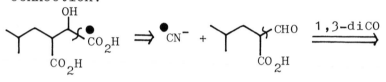

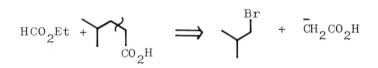

<u>Synthesis</u>: (<u>J. Amer. Chem. Soc.</u>, 1976, <u>98</u>,
6380; <u>Tetrahedron</u>, 1972, <u>28</u>, 1995).

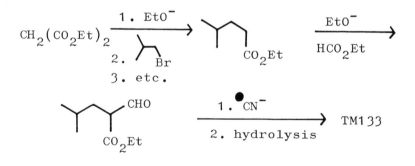

135. Here is a more difficult example based also on α-hydroxy acids. Use the two phenyl groups as a clue for your first disconnection in designing a synthesis for TM135:

TM135

136. <u>Analysis</u>: Using the clue, we remove both phenyl groups to give an ester:

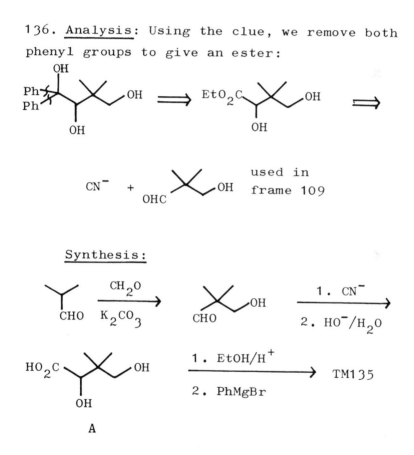

used in frame 109

Synthesis:

A

We can either protect the two hydroxyl groups in A as a cyclic acetal or use four mols of PhMgBr and waste two of them.

137. A more elaborate variation gives a general amino acid synthesis. If the reaction between an aldehyde and cyanide is done in the presence of ammonia, the product is an α-amino-nitrile:

$$RCHO \xrightarrow{\quad NH_3, \ CN^- \quad} \underset{RCH-CN}{\overset{NH_2}{|}}$$

Can you see what intermediate is being trapped by the cyanide ion?

138. It must be the imine:

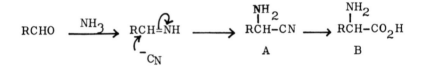

Under the right conditions, hydrolysis of the cyanide A occurs during the reaction to give the amino acid B. How could you make the amino acid Valine (TM138)?

TM138

139. Analysis:

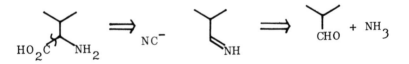

Synthesis:

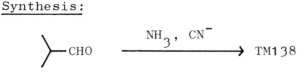

$$NH_3, \ CN^-$$

⟩—CHO ——————————→ TM138

This is the Strecker amino acid synthesis.

---

140. Strangely enough, cyanide ion is also
involved in one special reaction giving an α-
hydroxy-ketone. Can you show how the adduct
A of benzaldehyde and cyanide ion can give
a stable 'carbanion'?

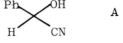

A

141.

   ——————→

This anion now reacts with another molecule
of benzaldehyde to give eventually the α-
hydroxy-ketone 141A. Draw mechanisms for
these steps:

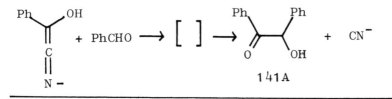

141A

142.

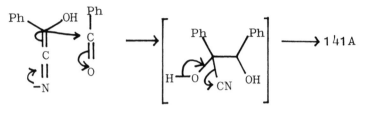

The product is called benzoin and the reaction is known therefore as the benzoin condensation.   No base is needed other than
cyanide ion.

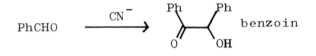

143. How could benzoin be elaborated into the

more complex molecule
TM143?

TM143

144. Analysis: We can disconnect both the
symmetrical α,β-unsaturated carbonyl linkages:

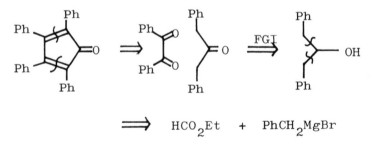

$$\implies \quad HCO_2Et \quad + \quad PhCH_2MgBr$$

Synthesis:

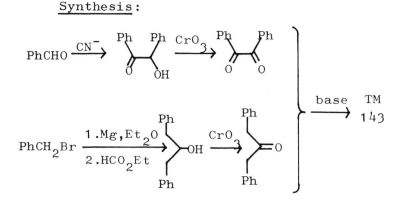

145. The same problem of illogicality arises
with other α-hydroxyketones:

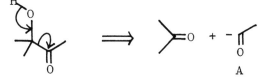

Again we need a reagent for an acyl anion
synthon (A). We find this in the acetylide
ion since substituted acetylenes can be
hydrated to ketones:

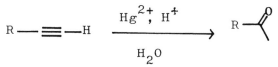

If you want to know more about this reaction,
see Norman p.116 or Tedder, vol 1, p.108.
How then could one
make TM145?

TM145

146. <u>Synthesis</u>:

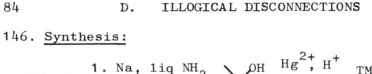

The reaction can be used for disubstituted acetylenes, but it is unambiguous only when they are symmetrical. Suggest a synthesis for TM146.

TM146

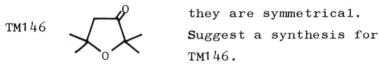

147. <u>Analysis</u>: The cyclic ether is obviously made from a diol, and that gives us a 1,2-dioxygenated skeleton of the right kind:

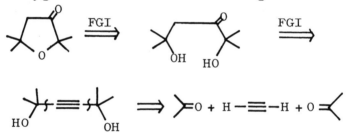

<u>Synthesis</u>: We need the symmetrical double adduct from acetone and acetylene.

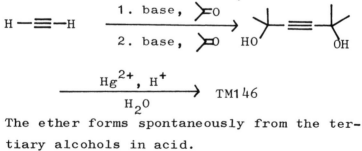

The ether forms spontaneously from the tertiary alcohols in acid.

148. α-Hydroxy ketones take part in condensation reactions
too.   How would   TM148
you make TM148?

149. <u>Analysis</u>: Start with the α,β-unsaturated
relationship as the alternative (the 1,2-diO)
is no good at the start.   After the first disconnection we have a methyl ketone which can
come from an acetylene:

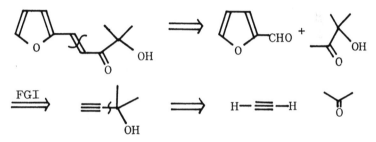

<u>Synthesis</u>: (<u>Ber.</u>, 1922, <u>55</u>, 2903 for the
later stages).

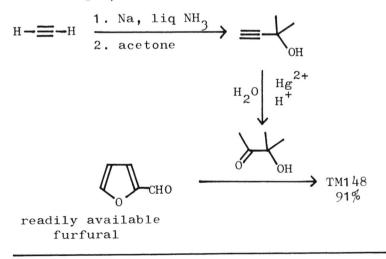

(b)   1,2-DIOLS

150. A good approach to 1,2-diols is the
hydroxylation of an olefin with reagents such
as OsO_4 or $KMnO_4$.   The olefin can be made by
the Wittig reaction so the disconnections are:

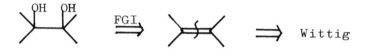

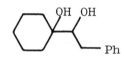

How could you make
this diol (TM150)?

TM150

---

151. Analysis: Back to the olefin - either
Wittig is possible but one pair of starting
materials is more readily available:

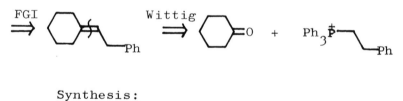

Synthesis:

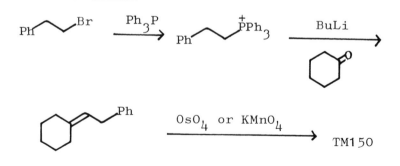

152. This hydroxylation gives <u>cis</u> addition to the double bond (see Norman p.502-3 or Tedder vol. 1, p.61-2 if you're not sure about this). How then could you make TM152?

TM152

153. <u>Analysis</u>: The acetal FG had better be removed first:

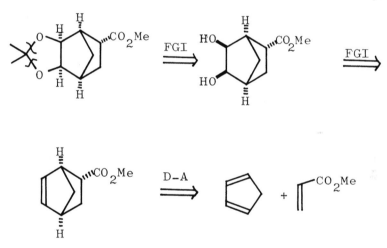

<u>Synthesis</u>: Assuming the usual stereo-selectivity for the Diels-Alder reaction and for the hydroxylation:

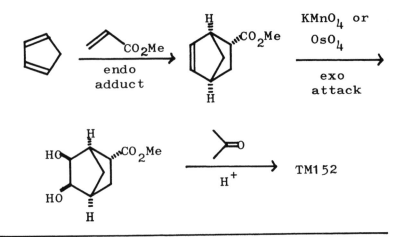

154. Symmetrical diols can be made by a radical reaction. Radical reactions are rarely much use in carbon-carbon bond formation as they often give poor yields and many products. They are of course useful in some FGI reactions in things like allylic bromination and in functionalising remote carbon atoms. If you want to read more about this see Tedder, Part 2, Chapter 11 or Carruthers, Chapter 4. One useful radical reaction is the pinacol reduction:

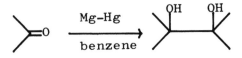

The disconnection is obvious, and gives us one way of making symmetrical 1,2-diols. What makes it more than trivial is that the products undergo the pinacol rearrangement:

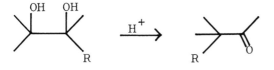

(see Tedder, Part 2, pp. 112-118, Norman,
p.438-440 if you aren't familiar with this
reaction).
How could you use this
route to make TM154?

TM154

155. Analysis: This is a t-alkyl ketone so a
pinacol rearrangement route will be possible:

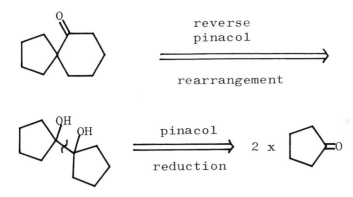

reverse
pinacol

rearrangement

pinacol

reduction

2 x

Synthesis: Since the pinacol is symmetrical, there is no ambiguity.

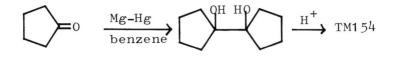

Mg-Hg

benzene

H$^+$

TM154

156. A closely allied reductive linking of carbonyl groups is an intramolecular version with esters, called the acyloin reaction, which again gives a 1,2-dioxygenated skeleton:

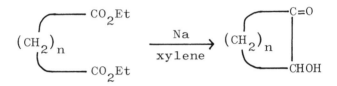

(details of the mechanism appear in Tedder, Part 3, p.193-4, or Norman, p.486-7). How could you make TM156?

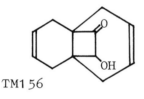

TM156

---

157. <u>Analysis:</u> First disconnect the acyloin product and the result is clearly made by D-A.

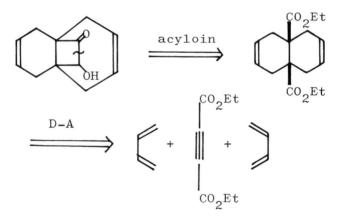

Synthesis: The conditions actually used
were (J. Org. Chem., 1966, 31, 2017):

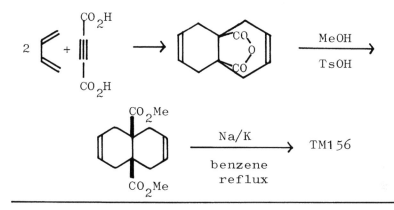

(c)   'ILLOGICAL' ELECTROPHILES

158. So far we have used 'illogical' nucleo-
philes and special methods to get round the
difficulty of making 1,2-dioxygenated compounds. Another approach is obviously to use
an 'illogicial' electrophile and among the
most important of these are α-halo-carbonyl
compounds. We can make these easily from
carbonyl compounds by halogenation of the enol
(see frames 198-207 of the Carbonyl Programme).

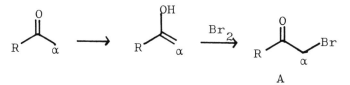

The enol is nucleophilic at the α carbon atom
but the α-bromoketone A is electrophilic at
the α carbon atom: by halogenation we have

inverted the natural polarity of the molecule.
How could you make TM158?

TM158
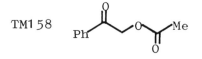

---

159. <u>Analysis:</u> As usual remove the ester
first:

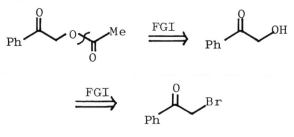

Synthesis: Since α-halo-carbonyl compounds are very reactive electrophiles, we can
use a short cut:

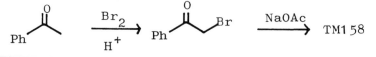

---

160. The hormone weed-killer MCPA (TM160) is
needed in large quantities. Suggest an

TM160   economical synthesis
             for it.

---

161. <u>Analysis:</u> The ether linkage can be disconnected directly on the alkyl side:

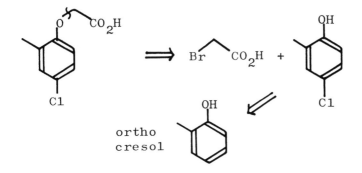

ortho
cresol

<u>Synthesis:</u> Chlorine is the cheapest of
the halogens, so it will be better to use
chloroacetic acid:

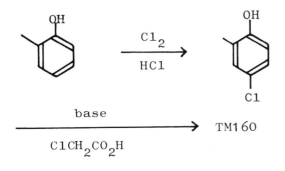

162. The other main illogical electrophiles
are epoxides, easily made from an olefin and a
per-acid, the usual one being m-chloroper-
benzoic acid (MCPBA) a commercial product.   A
more detailed explanation comes later, in
frames 276-7.

What product would you get from this and
sodium methoxide in methanol?

163.

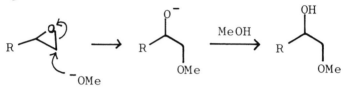

In acid solution, the other regio isomer would
be formed because the transition state (163A)
has a partial positive change on carbon
stabilised by R:

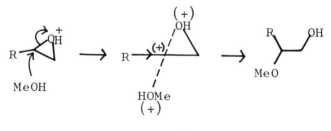

A

How would you make TM163 from simple starting materials?

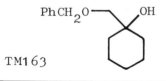

TM163

164. <u>Analysis:</u> This is the substitution
pattern for a base-catalysed epoxide opening:

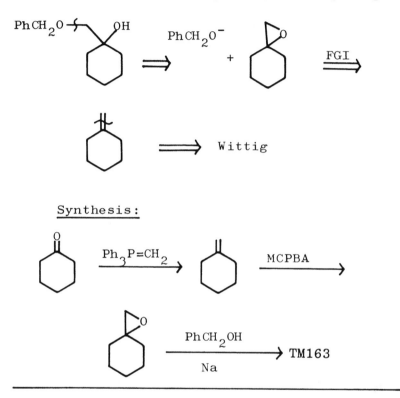

<u>Synthesis:</u>

165. Amines also react with epoxides at the
less substituted carbon atom. As a slightly
more testing problem, suggest a synthesis of
the alcohol (TM165) whose derivatives are
used in disinfectants ("phemeride" etc.).

166. <u>Analysis</u>: There is a series of 1,2
relationships here: it's easiest to start
with the free hydroxyl group:

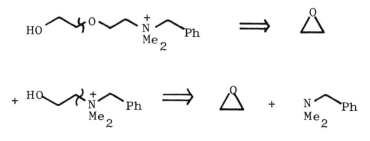

Synthesis: It might be better to add the
benzyl group at the end so that we can use
dimethylamine.

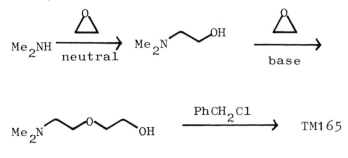

In base the OH becomes O⁻ and is more nucleophilic than N.   In neutral solution N is more
nucleophilic than OH.

(d)   <u>REVIEW PROBLEMS</u>

167. <u>Review Problem 12</u>: Suggest a synthesis
of the lactone acid
TM167.

TM167

168. Analysis: Opening the lactone reveals
1,2-, a 1,5-, and a 1,6- di-oxygenation
relationships. We must tackle the 1,2 first:

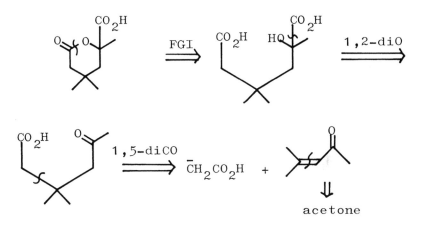

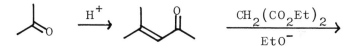

Synthesis: The first stages are well
known and the two methyl groups assist the
final cyclisation:

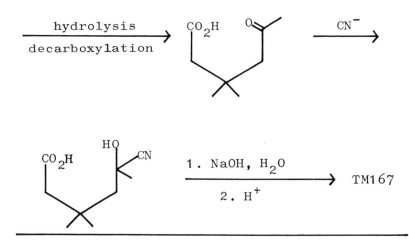

169. <u>Review Problem 13</u>: This odd looking
molecule (TM169) is closely related to multi-
striatin, a phenomone of the elm bark beetle,
the insect which spreads Dutch elm disease.
How would you synthesise a sample for testing on the beetle?

TM169

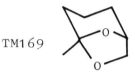

---

170. <u>Analysis</u>: The functional group is an
acetal - once we've removed this, we can
follow straightforward tactics.

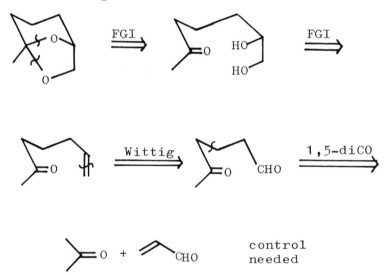

control
needed

<u>Synthesis</u>: We must be able to do a Wittig
reaction on the aldehyde but not on the ketone,
so we must protect the ketone: therefore add
the aldehyde as an ester (there are many other

solutions).

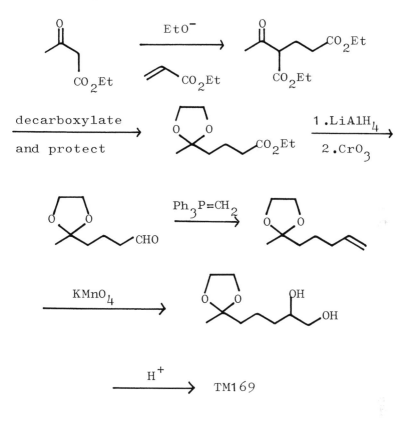

Cyclisation occurs by trans-acetalisation.

2.   THE 1,4-DIOXYGENATION PATTERN
(a)   1,4-DICARBONYL COMPOUNDS

171. The obvious disconnection on a 1,4-
dicarbonyl compound gives us a logical nucleophilic synthon (an enolate anion) A but an
'illogical' electrophilic synthon B:

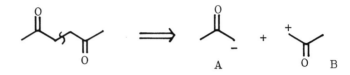

We need for B a derivative of a ketone in
which the normal polarity is inverted, and
you will realise from frames 158-161 that the
α-halo carbonyl compound is ideal. So how
would you make TM171?

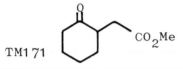

TM171

172. <u>Analysis</u>: Either way round will do, so
let's arbitrarily chose ketone and   α-halo
ester:

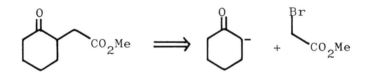

<u>Synthesis</u>: Our problems are not yet over
because if we combine ketone and α-halo ester
in base, quite a different reaction occurs.
Can you draw a mechanism for it?

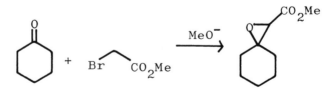

173. The most acidic proton is that next to
the ester and the halide:

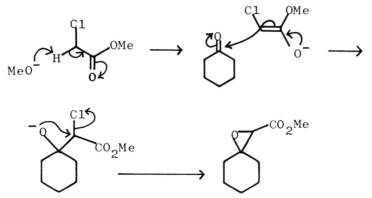

This, the Darzens reaction, is useful in
other circumstances (frames 280-1) but a
nuisance here. We must use some means to
make the ketone act as the nucleophile in the
initial condensation. One effective way is
to convert it into an enamine. Draw a mechanism for

this reaction.

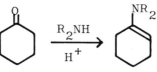

174.

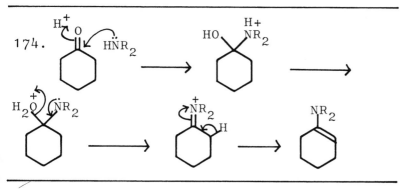

175. The complete synthesis of TM171 now becomes:

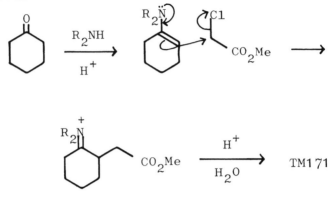

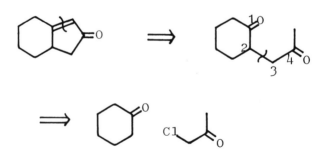

The enamine attacks the reactive α-carbonyl halide rather than the carbonyl group itself.

How could you make TM175?

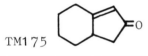

TM175

---

176. <u>Analysis:</u> Starting with the α,β-unsaturated ketone, we then have a 1,4-di ketone so we shall have to use our new method.

Synthesis: Actually done this way (J. Amer. Chem. Soc., 1958, 80, 6609):

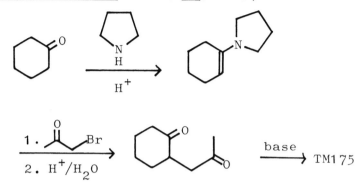

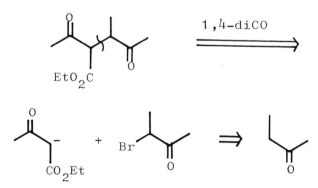

177. Enamines are not always necessary - sometimes the enolate anion is stable enough by itself. How would you make TM177?

$$\text{TM177}$$

178. Analysis: The activating group gives us a guide:

1,4-diCO

Synthesis: Bromination of butanone in
acid gives predominantly the isomer we want,
(see frames 198-201 of the Carbonyl Programme)
and again the reactive α-carbonyl halide is a
good electrophile:

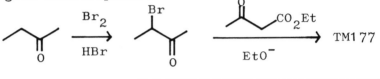

$$TM177$$

(b)   γ-HYDROXY CARBONYL COMPOUNDS

179. The α-halo carbonyl compounds are
reagents for the synthon $\overset{+}{C}-\overset{\overset{O}{\|}}{C}$. At a lower
oxidation level, epoxides are reagents for
the synthons $\overset{+}{C}-\overset{\overset{OH}{\mid}}{C}$ as in their reaction with
Grignard or organolithium reagents (see also
frames 162-6).

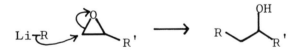

How might you synthesise TM179?

TM179

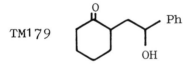

180. Analysis:

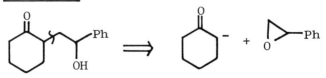

Synthesis: Again the enamine can be used to provide the enolate synthon:

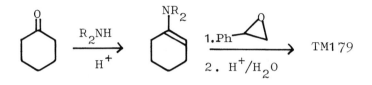

181. One particular sequence under this general heading is rather important. How would you make TM181?

TM181

182. Analysis: The usual disconnection gives a stable anion:

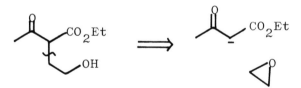

Synthesis: In fact this combination of reagents doesn't give TM181: instead the lactone 182A is formed. This lactone is useful in all the reactions for which we might plan to use TM181.

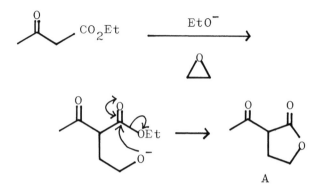

183. How then would you make the γ-halo

ketone TM183?

TM183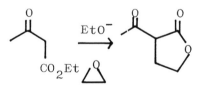

184. Analysis: The normal FGI gives the
alcohol, and disconnection of this in the
usual way gives:

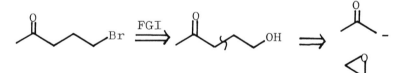

          Synthesis: Using CO_2Et as the activa-
ting group:

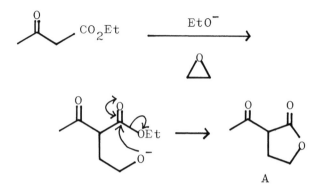

	1.hydrolysis	
	2.heat $(-CO_2)$	→ TM183
	3. OH → Br	

It turns out that the last three steps can be accomplished simply by boiling with conc. HBr.

---

185. How would you make this γ-halo ketone?

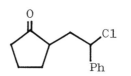

TM185

---

186. <u>Analysis:</u>

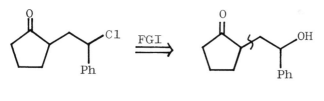

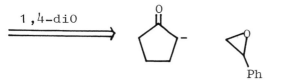

<u>Synthesis:</u> We shall need an activating group, and our starting material is actually TM104 !

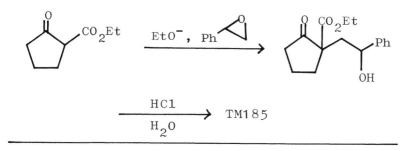

(c) <u>OTHER 'ILLOGICAL' SYNTHONS</u>

187. Clearly other combinations of logical
and illogical synthons could be used to make
1,4-dioxygenated compounds. How could you
use cyanide ion (as the $^-CO_2H$ synthon) to
make a   γ-keto acid such as

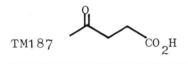

TM187

---

188. All we have to find is the electrophile:

How could you make the same acid using
propargyl bromide $BrCH_2C{\equiv}CH$ as your illogical
fragment?

---

189. Acetylenes can be hydrated to give
ketones (frame 145) so propargyl bromide must
provide a $MeCOCH_2^+$ synthon:

Synthesis: Activation will be needed:

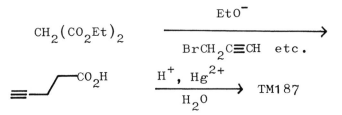

$$CH_2(CO_2Et)_2 \xrightarrow{\quad EtO^- \quad}_{BrCH_2C\equiv CH \ etc.}$$

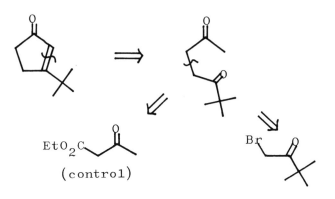

You will meet other combinations of logical and illogical synthons in review problems later.

(d)   REVIEW PROBLEMS

190. Review Problem 14: Cyclopentenones (e.g. TM190) occur in nature and are important in prostaglandin synthesis.
How would you make this   TM190
one?

---

191. Analysis: Start with the α,β-unsaturated carbonyl:

Synthesis: Ketone A is just pinacolone, the product of the pinacol rearrangement (frames 154-5).

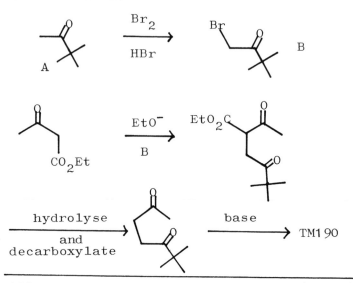

192. <u>Review Problem 15</u>: This triol (TM192) can be taken as a 1,4- or a 1,5-dioxygenated compound. In fact only one of these will work.

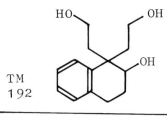

Suggest a synthesis.

TM 192

193. <u>Analysis</u>: We must go back to the corresponding tricarbonyl compound, writing CHO or CO_2Et for CH_2OH. Then we see that the 1,5-dicarbonyl relationship is no use as there isn't room for a double bond (e.g. 3-4) in the precursor we would have to write. We have to use the 1,4 relationships:

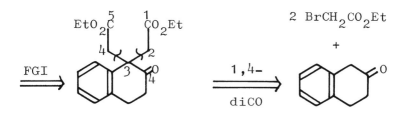

<u>Synthesis:</u> The ketone will enolise on
the side we want because of conjugation with
the benzene ring. It turns out that both
alkylations happen at once:

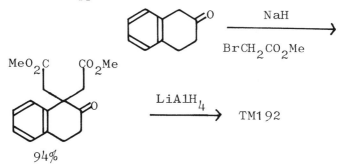

94%

The first reaction was involved in a synthesis
of morphine, the starting ketone being made
by reduction of a substituted naphthalene
(<u>J. Amer. Chem. Soc.</u>, 1950, <u>72</u>, 3704). No
doubt an epoxide could have been used as the
electrophile.

3.   1,6-<u>DICARBONYL COMPOUNDS</u>

194. Clearly, these too will be 'illogical
disconnections' but we can get round the problem in a different way by using a 'disconnection' which actually links up the two

carbonyl groups:

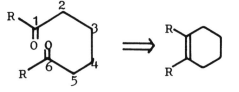

The reaction is
with ozone or its
equivalent

Try this for TM194, analysing its synthesis
back to simple
starting materials.

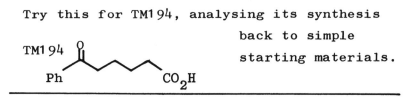

TM194

Ph   CO$_2$H

195. Analysis:

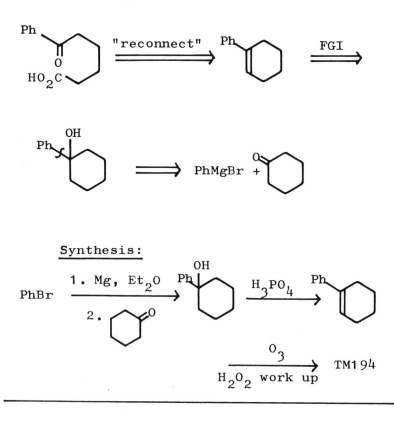

196. Since cyclohexenes can also be made by
the Diels-Alder reaction (frames 5-8) we have
access to a wide range of 1,6-dicarbonyl compounds.   How
about TM196?

HO_2C CO_2H

HO_2C — — CO_2H TM196

---

197. <u>Analysis</u>: Choosing the 1,6-dicarbonyl
relationship first:

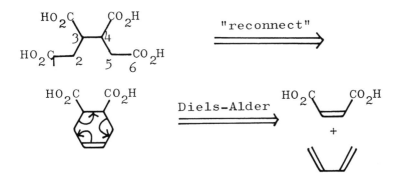

<u>Synthesis</u>: Maleic anhydride is the best
reagent.

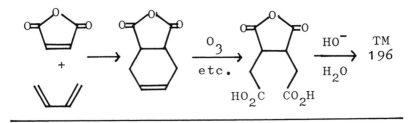

198. Another way to make cyclohexenes is by the partial reduction of benzene rings ('Birch reduction', described in Norman, p.553-557) such as:

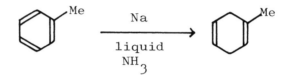

199. With that clue, how would you make:

TM199

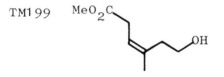

200. <u>Analysis</u>: First convert the 1,6-di-oxygenated compound to a 1,6-dicarbonyl compound, keeping the two carbonyl groups different:

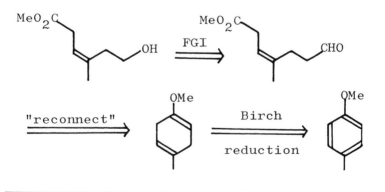

Synthesis: (E. J. Corey et al., J. Amer. Chem. Soc., 1968, 90, 5618).

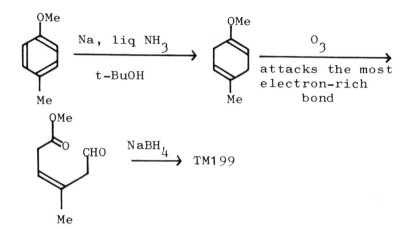

Note that the sterochemistry of the remaining double bond must be right as it has come from a ring.

---

201. Disconnection of other combinations of functional groups can lead us back to a 1,6-dicarbonyl compound.
Try this on TM201.   TM201

---

202. Analysis: Taking the α,β-unsaturated aldehyde first:

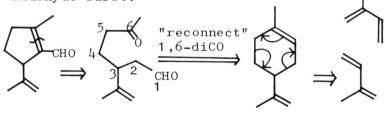

116   D.   ILLOGICAL DISCONNECTIONS

Synthesis: The Diels-Alder reaction is simply the dimerisation of isoprene to give the naturally occurring terpene A.   Now we have to cleave one double bond and leave the other alone.   It turns out that epoxidation is selective in this case.

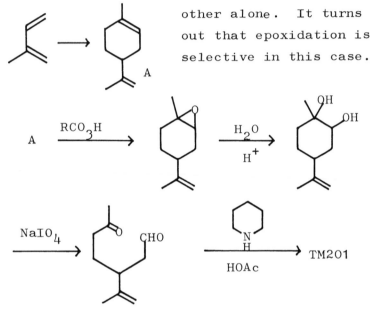

The condensation conditions must be as mild as possible, because we want to get only the most stable of the three possible enols (from the aldehyde).   Though you could not have predicted the exact conditions either for the double bond cleavage or for the condensation, you should have seen that control was possible as in each case the two functional groups are different enough.   (J. Amer. Chem. Soc., 1960, 82, 636; J. Org. Chem., 1964, 29, 3740; Tetrahedron Letters, 1965, 4097).

4.   REVIEW SECTION.
     SYNTHESIS OF LACTONES
     (This section may be worked now, or at
any later stage for revision).

203. We have now considered all the simple
two-group disconnections and you should
be able to design reasonable syntheses for
most small molecules. As a set of review
problems, try to design good syntheses for
these lactones. In each case only one or two
answers are suggested, but others will be as
good. Discuss your answers with someone else
if they are substantially different from the
suggestions.

___

204. Review Problem 16: Design a synthesis
for TM204, an intermediate in Khorana's
Coenzyme A Synthesis,
J. Amer. Chem. Soc.,   TM204
1961, 83, 663.

___

205. Analysis: First open the lactone to
reveal the true target.

Note the use of
CN^- as a reagent
for $^-CO_2H$.

Synthesis: A mild base must be used to avoid the Cannizzaro reaction. The hydroxy acid A cannot be isolated and cyclises spontaneously. (Fleming p.92).

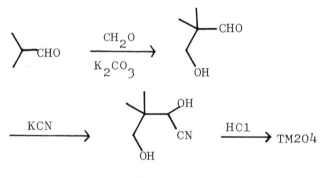

206. Review Problem 17: Design a synthesis for TM206, an intermediate in Crandall and Lewton's biogenetically patterned synthesis of cedrene (J. Amer. Chem. Soc., 1969, 91, 2127).

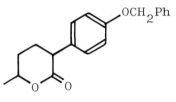

TM206

207. Analysis: This is a 1,5-dioxygenated skeleton, therefore further FGI is necessary to give a 1,5-dicarbonyl compound.

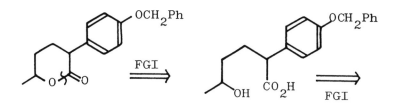

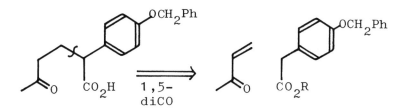

Synthesis: An activating group is necessary to control the Michael reaction:

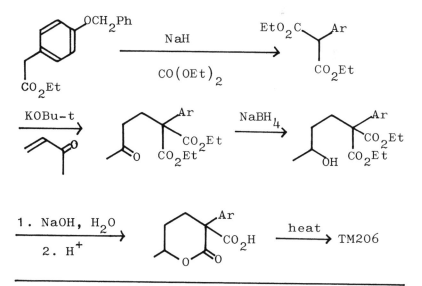

208. <u>Review Problem 18</u>: Design a synthesis for
TM208, an intermediate in Woodward's tetra-
cycline synthesis (<u>Pure and Applied Chemistry</u>,
1963, <u>6</u>, 561, <u>J. Amer. Chem. Soc.</u>, 1968, <u>90</u>,
439).

TM208

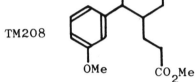

209. <u>Analysis</u>: The same FGI as in problem 5
gives us the key intermediate A. (frame 81)

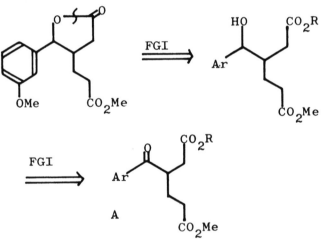

A contains 1,4-, 1,5-, and 1,6-dicarbonyl
relationships, so many disconnections are possible, summarised in the following chart.

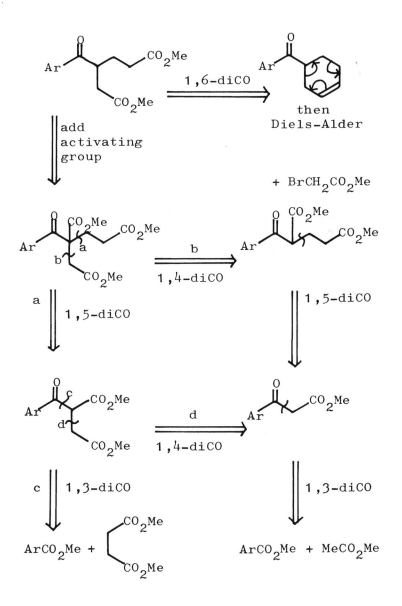

Woodward tried all these different routes
except the one based on the 1,6-dicarbonyl
relationship. All were successful, but he
eventually chose the route corresponding to
<u>a</u> and <u>c</u>. He discusses this synthesis at
length in the 1963 reference.

Synthesis: (Just this one route!) The
hydroxy acid in the final step could not be
isolated.

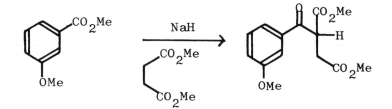

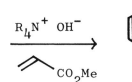 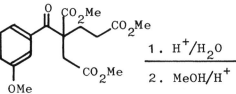

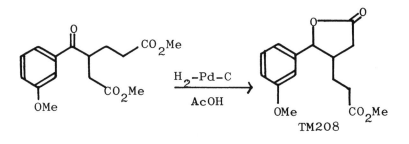

TM208

210. Now that you have finished the second
systematic section and are familiar with both
one-group and two-group disconnections, it is
important that you work through some general
problems without being told which particular
disconnection to do.   The problems are meant
to be graded in difficulty.   I suggest you do
as many as you need to convince yourself that
you can cope.   Later you can come back and do
the rest.

Review Problem 19: Design a synthesis
for TM210.

AcO   TM210

---

211. Analysis: The ester group is obviously
just FGI, but immediate disconnection of the
alcohol A doesn't get us very far so we do a
bit more FGI.   The double bond is the guide
as it can be added as an allyl group:

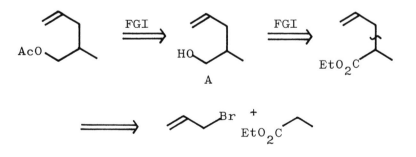

Synthesis: An activating group will be needed:

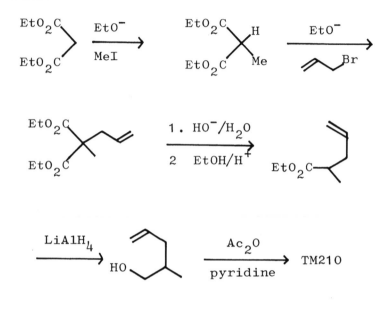

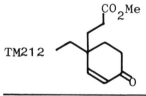

___

212. Review Problem 20: Suggest a synthesis for TM212, an intermediate in Stork's synthesis of the complex alkaloid aspidospermine. J. Amer. Chem. Soc., 1963, 85, 2872).

TM212

___

213. <u>Analysis:</u> Start with the only recognis-
ably helpful relationship, the α,β-unsatu-
rated ketone:

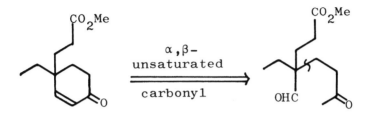

We now have two 1,5-diCO relationships: disconnect the more reactive one first.

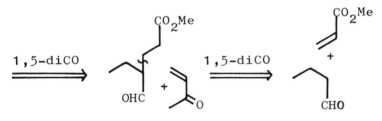

    <u>Synthesis:</u> We shall need the usual acti-
vating group for both Michael reactions: it
can't be a CO_2R group as there isn't room, so
it will have to be an enamine. The synthesis
is therefore:

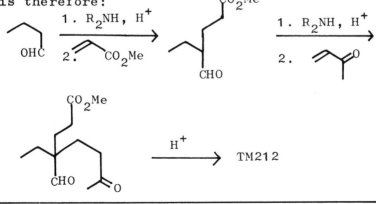

214. <u>Review Problem 21</u>: Design a synthesis
for TM214.

TM214

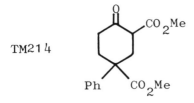

---

215. <u>Analysis</u>: You have two 1,5- and one 1,3-
dicarbonyl relationships to consider. Disconnections like   <u>a</u>   hardly simplify the problem
at all, whereas disconnecting the 1,3-di-
carbonyl relationship <u>b</u>   gives a symmetrical
intermediate:

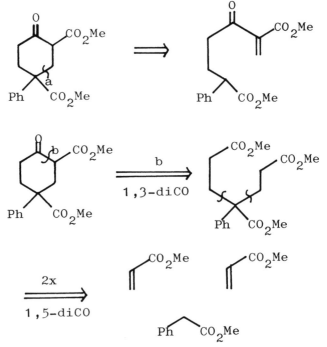

Synthesis:

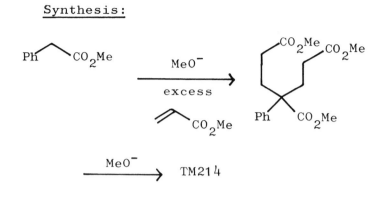

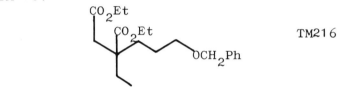

---

216. <u>Review Problem 22</u>: Design a synthesis
for TM216.

TM216

---

217. <u>Analysis:</u> This is part of a Kutney
alkaloid synthesis (<u>J. Amer. Chem. Soc.</u>, 1966,
<u>88</u>, 3656). There is a 1,4- and a 1,5-di-
oxygenation relationship; choosing the 1,4-
first as it is at the right oxidation level
we get:

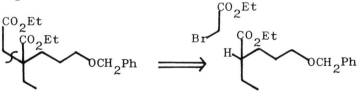

We shall need a strong base as there's no
room for an activating group.

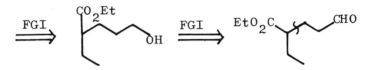

This is at the right oxidation level for a
1,5-diCO disconnection.

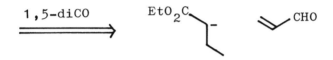

An activating group will be needed.

Synthesis: This was how Kutney did it —
there are other orders of events.

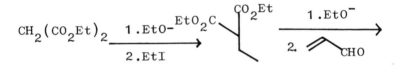

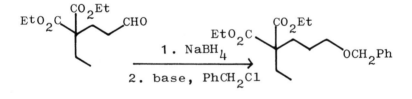

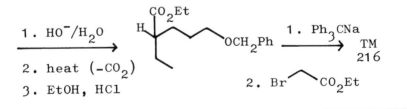

218. <u>Review Problem 23</u>: Design a synthesis
for TM218:

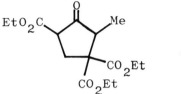

TM218

---

219. <u>Analysis</u>: This molecule has about every
relationship in the book!   One reasonable
answer is this:

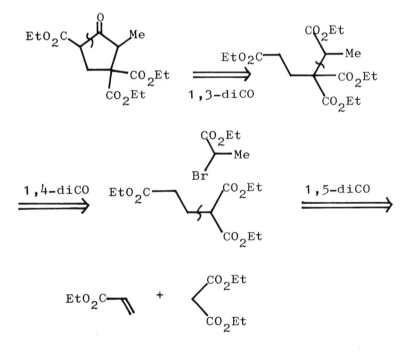

Synthesis: Russian workers actually carried out the synthesis in a different order; the logic is the same:

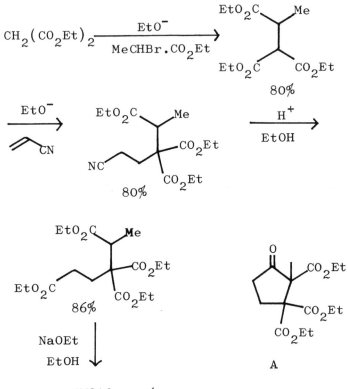

TM219, 91%

The alternative condensation to give A does not happen because A cannot form a stable enolate ion, whereas TM219 can. <u>Zhur. obshchei Khim.</u>, 1957, <u>27</u>, 742; <u>Chem. Abs.</u>, 1957, <u>51</u>, 16313.

220. The most important pericyclic reaction
in synthesis, indeed one of the most important of all synthetic methods, is the Diels-
Alder reaction.   We have seen this many times
before.   What are the clues for a Diels-Alder
disconnection?

---

221. A cyclohexene with an electron-withdrawing group on the other side of the ring to
the double bond:

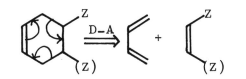

$Z = COR, CO_2Et,$
$CN, NO_2,$ etc.

So how would you make TM221?

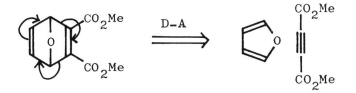

TM221

---

222. Simply reverse the Diels-Alder reaction:

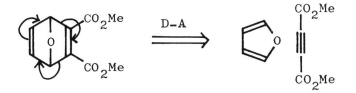

Both starting materials are readily available.
What about TM222?

TM222

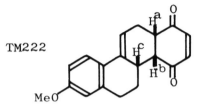

---

223. Again, simply reverse the Diels-Alder.
It may have taken you a little time to find
the right cyclohexene!

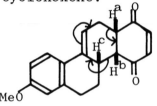

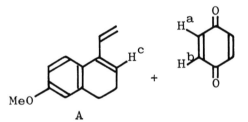

A

Note that the stereochemistry comes out right.
H's a and b are cis because they were cis in
the starting quinone and the Diels-Alder reaction is stereospecific in this respect. H^c is
also cis to H^a and H^b because the Diels-Alder
reaction is stereoselectively endo.   These
points are described in more detail in Norman
p.284-6 and explained in Ian Fleming 'Frontier
Orbitals and Organic Chemical Reactions',

Wiley 1976, p.106-109. How would you make diene A?

___

224. <u>Analysis:</u> We must put in a hydroxyl group instead of a double bond and the best place to do this is, as usual, at the branch point:

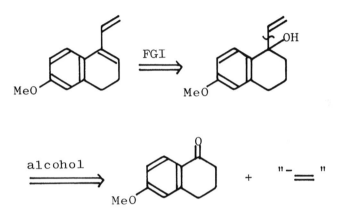

<u>Synthesis:</u> The vinyl anion synthon can either be the vinyl Grignard reagent or the acetylide anion, in which case the synthesis becomes:

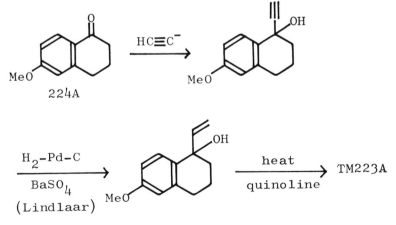

There are many syntheses of 224A which was a
famous intermediate in early steroid syntheses
(e.g. J. Amer. Chem. Soc., 1947, 69, 576,
2936),

---

225. Since the Diels-Alder reaction is so good
it's worth going to some trouble to get back
to a recognisable Diels-Alder product.   Take
TM225 for example.   The first D-A disconnection
is obvious, but can you find your way back to
a second?

TM225

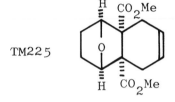

---

226. Analysis: The obvious D-A first!

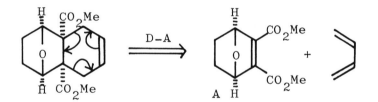

All we have to do for a second D-A disconnection is put another double bond into A in the
right position.

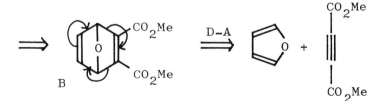

B is actually TM221

Synthesis:

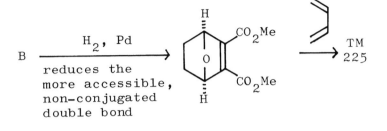

227. FGI of a different kind should give you
another D-A disconnection on TM227. Don't
forget the stereochemistry!

TM227

228. <u>Analysis:</u> We must first find some carbonyl groups somewhere:

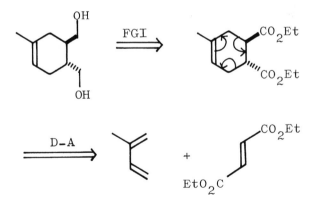

Synthesis: The ester must be a fumarate so that the stereochemistry of the final adduct is correct:

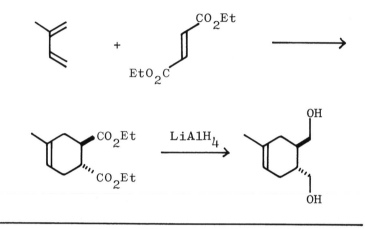

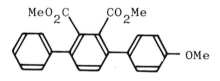

229. This next example is perhaps slightly less obvious - the problem of how to make an asymmetrically substituted terphenyl (e.g. TM229) but inspection should show you which ring is made by the D-A reaction.

TM229

230. Analysis: The central ring has the electron-withdrawing substituents so all we have to do is to adjust the oxidation level:

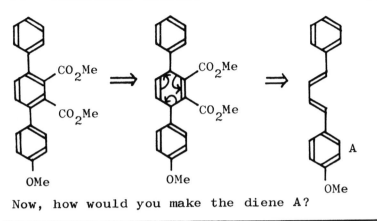

Now, how would you make the diene A?

---

231. <u>Analysis</u>: A Wittig disconnection would give two readily accessible fragments:

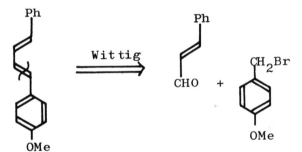

The aldehyde is the readily available cinnamaldehyde; the bromide can be made from <u>p</u>-cresol.

Synthesis:

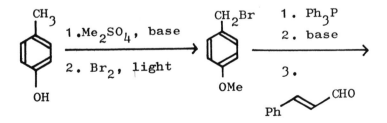

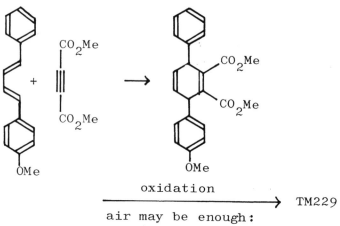

oxidation

air may be enough:
a quinone is often used

TM229

REVIEW PROBLEM

232. Review Problem 24: Design a synthesis
for TM232.

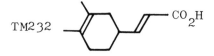

TM232

---

233. Analysis: Disconnect the α,β-unsaturated
acid first and then an obvious D-A is
revealed:

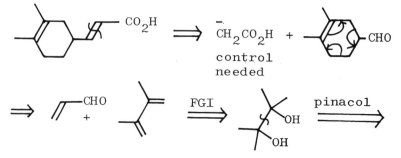

Synthesis: The conditions actually used
were:

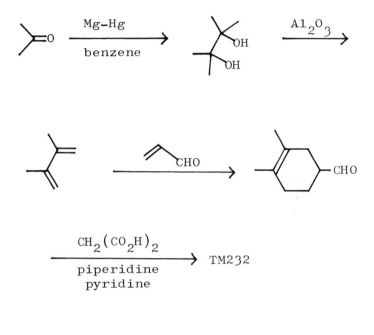

$$\xrightarrow[\substack{\text{piperidine} \\ \text{pyridine}}]{CH_2(CO_2H)_2} \quad TM232$$

1. HETEROATOMS, ETHERS, AND AMINES

234. Any heteroatom (usually O, N, or S) in
a carbon chain is a good point for a disconnection and I shall write CO, CN, or CS above
the arrow when I use these disconnections.
TM234 can be made
this way. How
would you actually
do it?

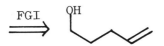 TM234

---

235. Analysis: We must choose the bond away
from the aromatic ring as displacements on
PhBr are almost impossible.

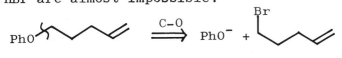

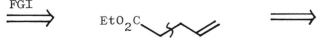

The double bond is so far away from the
hydroxyl group that we shall have to alter
the oxidation level before we can continue:

FGI
⟹ EtO_2C ⟍⟋⟍⟋   ⟹

        control   Br   reactive
                                  allyl
    $(EtO_2C)_2CH_2$ +   ⟍⟋   bromide

Synthesis:

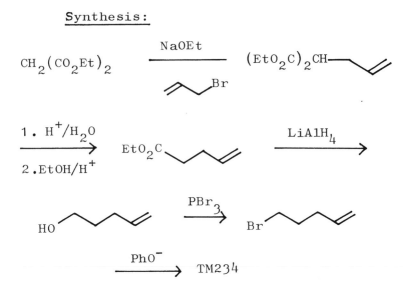

$$\text{CH}_2(\text{CO}_2\text{Et})_2 \xrightarrow[\text{Br}]{\text{NaOEt}} (\text{EtO}_2\text{C})_2\text{CH}-$$

$$\xrightarrow[\text{2.EtOH/H}^+]{\text{1. H}^+/\text{H}_2\text{O}} \text{EtO}_2\text{C} \xrightarrow{\text{LiAlH}_4}$$

$$\text{HO} \xrightarrow{\text{PBr}_3} \text{Br}$$

$$\xrightarrow{\text{PhO}^-} \text{TM234}$$

---

236. Amines are slightly more of a problem
because the same disconnection is no good:

$$\text{PhN} \atop \text{H} \centernot\Longrightarrow \text{PhNH}_2 + \text{Br}$$

Why is it no good?

---

237. Because it will be impossible to prevent
polyalkylation since the product is more
nucleophilic than the starting material:

$$\text{Ph}\ddot{\text{N}}\text{H}_2 \xrightarrow{\text{RBr}} \text{Ph}\ddot{\text{N}}\text{H}{-}\text{R} \xrightarrow{\text{RBr}} \text{PhNR}_2 \longrightarrow \text{etc.}$$

The trick used is to <u>acylate</u> the amine
instead, since we can reduce the resulting
amide with LiAlH$_4$ to give the product we want:

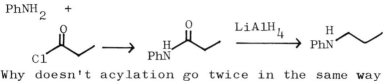

PhNH$_2$ +

Why doesn't acylation go twice in the same way as alkylation?

---

238. Because the acylated product has a delocalised lone pair and is less reactive than PhNH$_2$. You may have been surprised that LiAlH$_4$ reduction com-
pletely removes the
carbonyl oxygen atom.
To help explain this, please draw the likely intermediate.

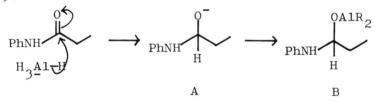

239.

A   B

The obvious intermediate, 239A, will now react with some aluminium species to give an inter-mediate like 239B, which can react further if the lone pair on nitrogen helps to expel the oxygen atom. Try now to complete the mechanism.

240.

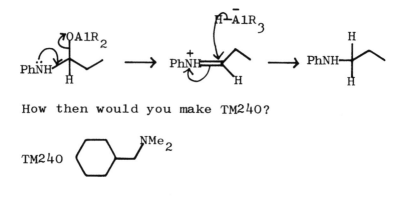

How then would you make TM240?

TM240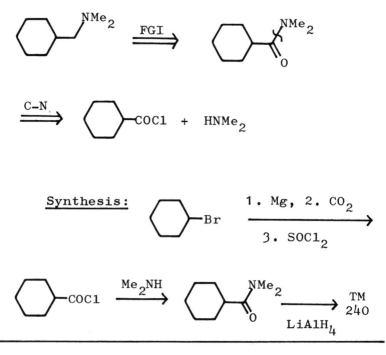

---

241. <u>Analysis:</u> The first step is to put in a carbonyl group next to nitrogen and then reverse the acylation:

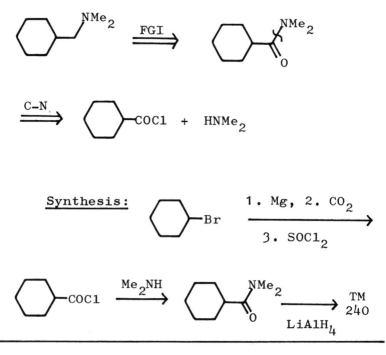

242.<u>Reduction</u> seems to be the keyword in amine synthesis since we can also reduce these functional groups to amines:

<u>Oximes</u>

<u>Nitriles</u>

<u>Nitro Compounds</u>

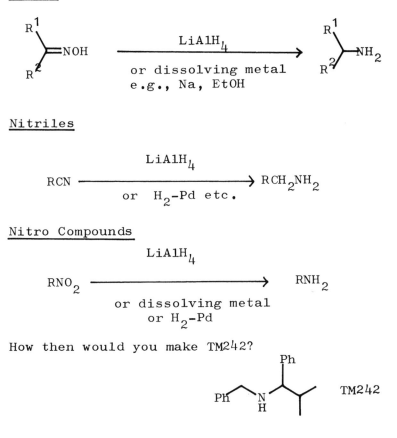

How then would you make TM242?

243. <u>Analysis:</u> The branched chain can only be made by reduction of an oxime so we must disconnect the other (benzyl) side first.

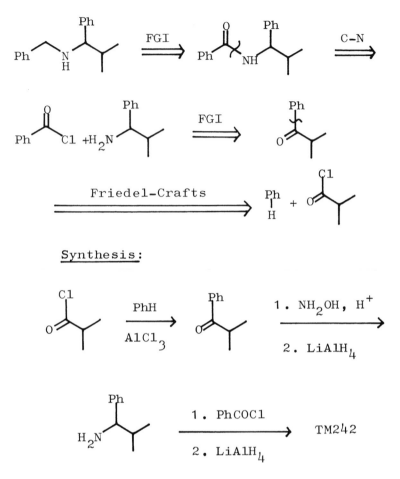

Synthesis:

1. NH$_2$OH, H$^+$

2. LiAlH$_4$

1. PhCOCl

2. LiAlH$_4$

TM242

244. 2-Arylethylamines (e.g. TM244) are
important intermediates in the synthesis of
alkaloids as you will see later. Suggest an
approach to TM244.

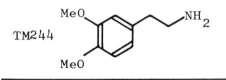

TM244

245. <u>Analysis:</u> There are two general ones
based on reduction of nitrile or nitro
compound:

(a)   <u>nitrile route</u>

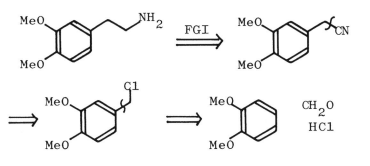

(b)   <u>nitro compound route</u>

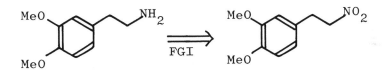

This compound could be easily made if it had
a double bond.   Since we are going to reduce
it anyway, this doesn't matter.

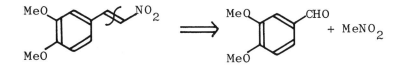

Using the same idea of reducing a nitro compound and a double bond at the same time, how

might we make TM245?

TM245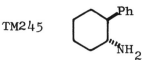

---

246. **Analysis:** The nitro compound looks like
a Diels-Alder adduct, so we know where to put
the double bond!

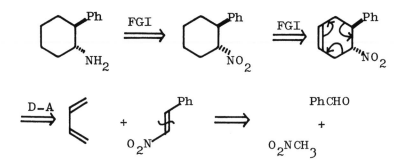

Synthesis: The trans nitro compound is
the one we get by condensation as it is more
stable than the cis compound.

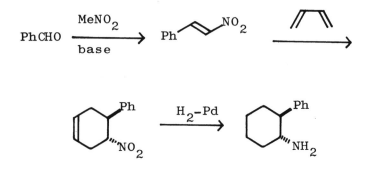

Now back to TM244.   How could you develop it
into TM246?

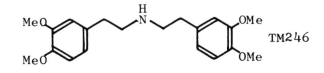

TM246

---

247. <u>Analysis:</u> We must put in a carbonyl group
again:

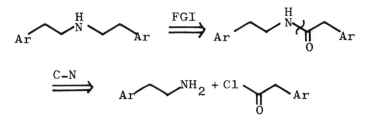

<u>Synthesis:</u> The most economical route will
therefore be to make both the acid chloride
and TM244 from the nitrile: (see Norman,
p.614-5):. (see also frame 45)

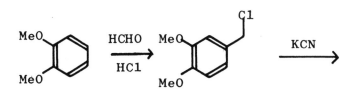

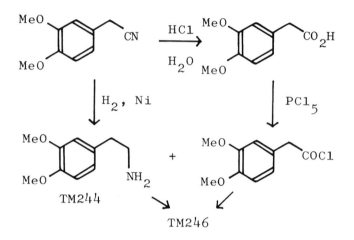

TM244   +   TM246

---

2.   HETEROCYCLIC COMPOUNDS

248. <u>Intramolecular</u> reactions are faster and
cleaner than <u>intermolecular</u> reactions. When
we want to make a C-N bond in a ring, therefore, we no longer have to take any special
precautions and we can use a nitrogen nucleophile and any carbon electrophile. Two useful disconnections are:

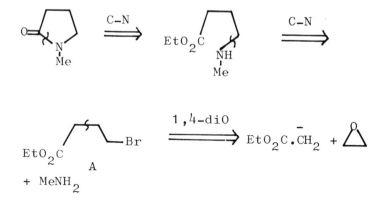

Simply mixing MeNH$_2$ and the γ-bromoester A
will give the heterocycle in one step.   How
might you
make TM248?

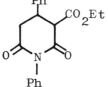

TM248

---

249. <u>Analysis</u>:   It is quicker to disconnect
both C-N bonds at once:

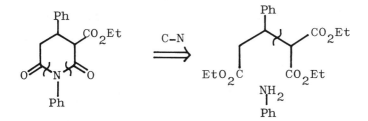

This is now a familiar 1,5-dicarbonyl
problem: the extra CO$_2$Et group tells us
where to disconnect.

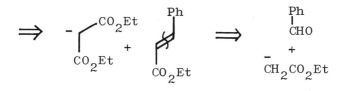

Synthesis:

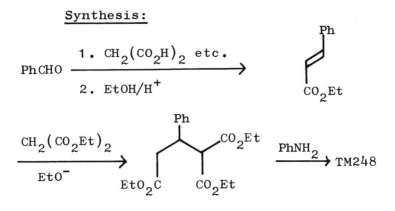

PhCHO $\xrightarrow{\begin{array}{c}1.\ CH_2(CO_2H)_2\ etc.\\ \hline 2.\ EtOH/H^+\end{array}}$

$\xrightarrow{\begin{array}{c}CH_2(CO_2Et)_2\\ \hline EtO^-\end{array}}$ $\xrightarrow{PhNH_2}$ TM248

The next example uses another carbon electrophile: how can you use the relationship of
the two functional groups in TM249 to design
a synthesis of the molecule?

TM249
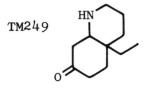

---

250. <u>Analysis</u>: The electrophile is an enone
since a reverse Michael reaction cleaves the
C–N bond:

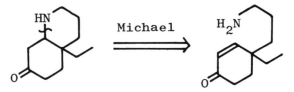

The long-chain amine can most quickly be
reduced to size via the nitrile.

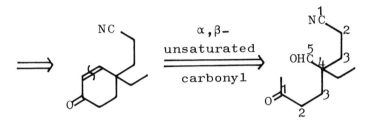

We now have two 1,5-diCO relationships which
we can disconnect in any order.

Synthesis: The final stages are very
similar to Review Problem 20 (frames 212-3).
TM249 is an intermediate in Stork's synthesis
of Aspidosperma alkaloids.   Stork's method
was actually a variation on the one we have
proposed (J. Amer. Chem. Soc., 1963, 85, 2872):

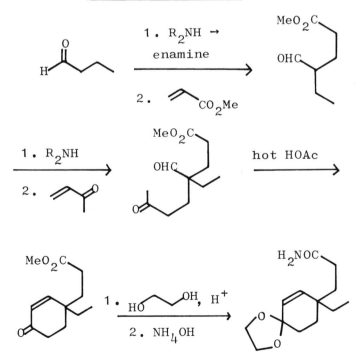

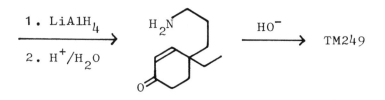

251. A more complicated structure doesn't always mean a more complicated disconnection when rings are being formed. Design a synthesis for TM251.

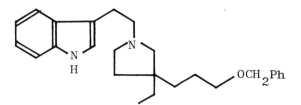

TM251

Concentrate on the saturated five-membered ring part first.

252. <u>Analysis</u>: We are clearly going to put carbonyl groups on each side of the five membered ring and disconnect the bond: e.g.

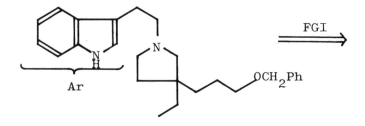

G.   HETEROATOMS   155

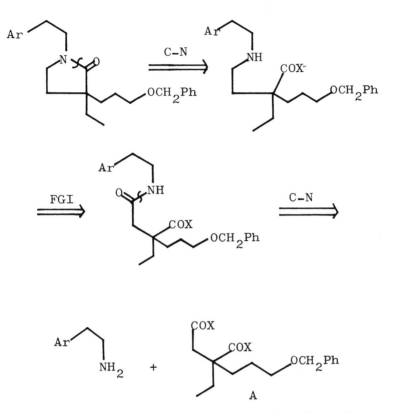

The analysis of intermediate A is given in
frame 217.

   Synthesis: This is part of Kutney's
quebrachamine synthesis (J. Amer. Chem. Soc.,
1966, 88, 3656). He found that if X = OEt,
both C-N bond-forming operations could be
carried out in one step. The synthesis is
actually much easier to carry out than
expected!

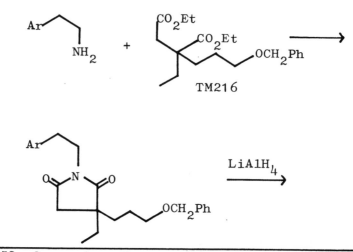

253. One extra disconnection is all we need
to cope with unsaturated heterocycles. If a
nitrogen atom is joined to a double bond in a
ring, we have a cyclic enamine. This is made
from an amine and a carbonyl compound in the
same way as ordinary enamines:

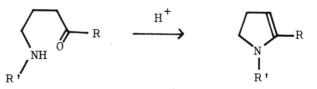

Draw a mechanism for this reaction.

254.

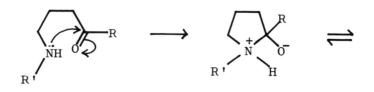

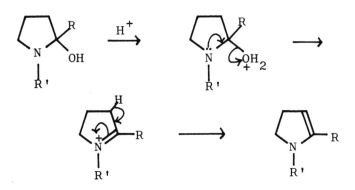

The disconnection corresponding to this
reaction is again of the C-N bond, writing an
amine and a carbonyl group in the right places

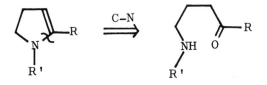

How then would you make TM254?

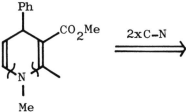

TM254

---

255. <u>Analysis:</u> Using the disconnection we've
just learned:

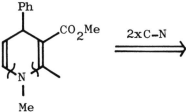

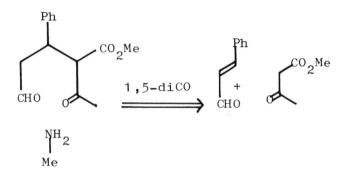

Synthesis: (As carried out in J. Amer. Chem. Soc., 1976, 98, 6650).

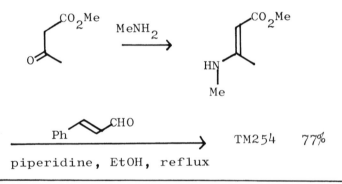

MeNH$_2$

piperidine, EtOH, reflux ⟶ TM254   77%

256. Many unsaturated heterocycles are made directly from dicarbonyl compounds. How would you make TM256?

TM256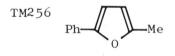

257. Analysis:

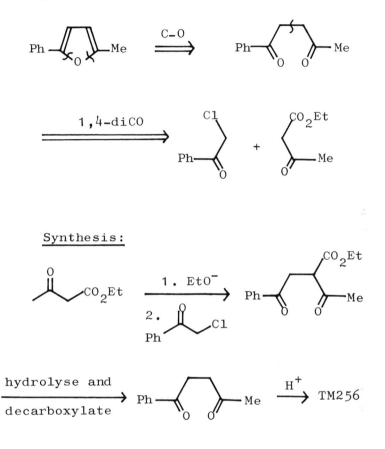

Synthesis:

hydrolyse and decarboxylate ⟶ Ph ⟶ Me $\xrightarrow{H^+}$ TM256

258. Heterocycles with two heteroatoms present no special problems and there are often several ways to do the first disconnection. One guide is to look for a small recognisable fragment containing the two heteroatoms. Try this one:

TM258

259. <u>Analysis</u>: The usual disconnections give us NH_2NH_2 (hydrazine) immediately:

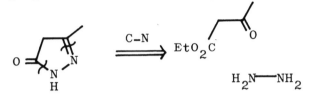

The synthesis is just to mix these two.

---

260. Here is one where the heteroatoms are apart:

TM260

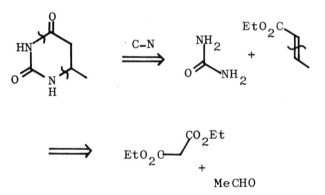

---

261. <u>Analysis</u>: Again the usual disconnection, making sure we choose a simple electrophilic fragment:

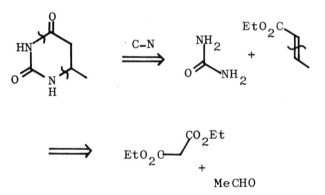

Synthesis: The heteroatom fragment is
urea.

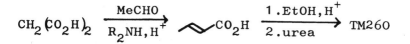

$$CH_2(CO_2H)_2 \xrightarrow[R_2NH,H^+]{MeCHO} \sim\sim CO_2H \xrightarrow[2.urea]{1.EtOH,H^+} TM260$$

262. So, to summarise this last section, disconnections of heterocycles are usually good
providing one checks the oxidation level:

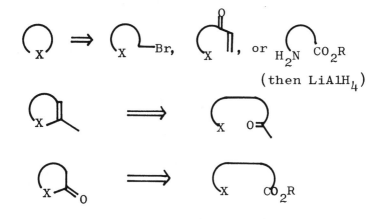

(then LiAlH$_4$)

The heteroatom is the nucleophile in all
these reactions: you just have to choose the
right electrophile.

263. There are many special methods to making heterocycles: if you want to read about
them, see Tedder, part 3, pp.115-131 and
205-220, or Norman, Chapter 18, p.588. We
are more interested in applying these general
methods to molecules in which a heterocyclic
ring is only part of the problem. How would

you make TM263 from simple starting materials?

TM263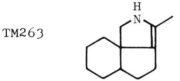

---

264. <u>Analysis:</u> The first disconnection is
easy:

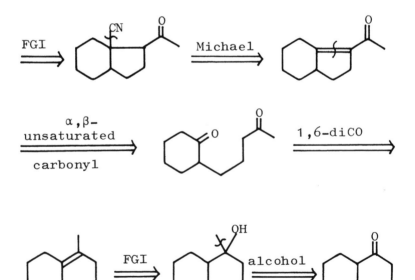

now if we make the primary amine from the
nitrile, we shall have another good disconnection.

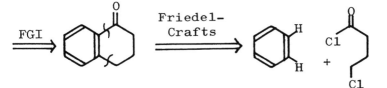

Synthesis: The early stages of this route are a well established method to make benzo-cyclohexanones; the later stages are adapted from Johnson's Conessine synthesis (J. Amer. Chem. Soc., 1962, 84, 1485):

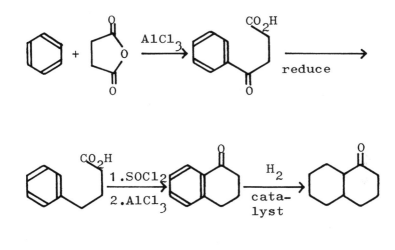

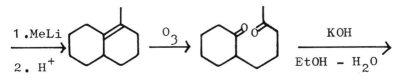

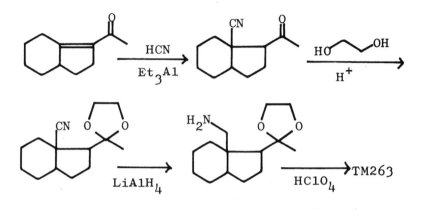

---

3.   AMINO ACIDS

265. You will have noticed that, throughout
this chapter, the heteroatom has always been
the nucleophile.   There is one way to use
nitrogen as an electrophile however and this
provides a good synthon for amino acid synthesis:

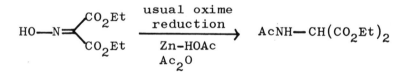

$$CH_2(CO_2Et)_2 \xrightarrow[(=NO^+)]{HONO} H^+ \quad O=N-\overset{H}{C}(CO_2Et)_2 \longrightarrow$$

$$HO-N=\begin{matrix} CO_2Et \\ CO_2Et \end{matrix} \xrightarrow[\substack{Zn-HOAc \\ Ac_2O}]{\substack{usual\ oxime \\ reduction}} AcNH-CH(CO_2Et)_2$$

$$\xrightarrow{base} AcNH-\overset{-}{C}=(CO_2Et)_2$$
$$\qquad\qquad\quad A$$

The anion A is a reagent for the synthon
$H_2N\text{-}\overset{\_}{C}H\text{-}CO_2H$ and amino acids are derived this
way. How could you make TM265?

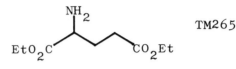

TM265

---

266. <u>Analysis</u>: Using the reagent we've just
made, it's easy to see what the electrophile
must be:

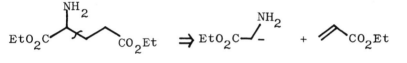

      <u>Synthesis</u>:

$$CH_2(CO_2Et)_2 \xrightarrow[\text{2.Zn-HOAc-Ac}_2\text{O}]{\text{1.HONO}} AcNHCH(CO_2Et)_2$$

$$\xrightarrow[\text{2.} \diagup CO_2Et]{\text{1.EtO}^-} AcNH\text{-}\underset{\diagdown CO_2Et}{\overset{|}{C}}(CO_2Et)_2 \xrightarrow[\text{2.H}^+/\text{EtOH}]{\text{1.base hydrolysis}}$$

    TM265

---

4.   <u>REVIEW PROBLEMS</u>

267. Before we leave heterocycles and heteroatoms, here are three review problems to rein-
force the ideas from this chapter. The first
two involve sulphur: don't be put off by that,

simply treat it as a special kind of oxygen.

   Review Problem 25: Design a synthesis
for TM267.

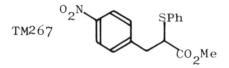

TM267

---

268. Analysis: As usual (see frame 235) we
shall disconnect the alkyl-heteroatom bond:

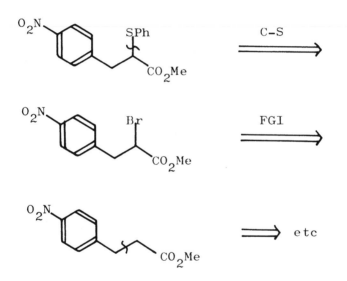

Synthesis: PhSH is readily available.

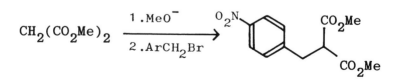

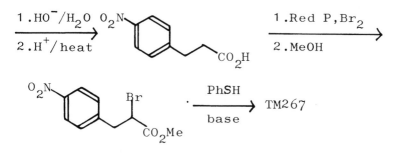

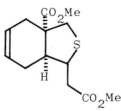

---

269. <u>Review Problem 26</u>: Design a synthesis
for TM269.

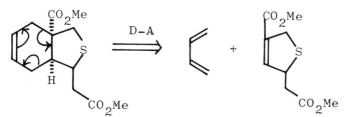

TM269

---

270. <u>Analysis</u>: We have an obvious Diels-Alder
disconnection, some C-S bonds, and a 1,6-
dicarbonyl relationship. The only one that
gives any rapid simplification is the D-A,
so we'll start with that:

There's still no helpful C-S disconnection,
so let's do the α,β-unsaturated ester next.

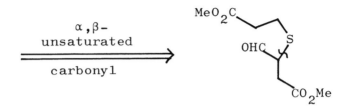

$$\xleftarrow{\begin{array}{c}\alpha,\beta- \\ \text{unsaturated} \\ \hline \text{carbonyl}\end{array}}$$

Both C-S bonds are now β to carbonyl groups and so can be disconnected in turn by reverse Michael reactions.

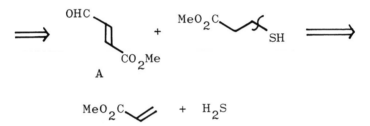

Synthesis: Fragment A is going to be very difficult: it would be much simpler to make it a diester and adjust the oxidation level later. This is the synthesis actually used by Stork (J. Amer. Chem. Soc., 1969, 91, 7780):

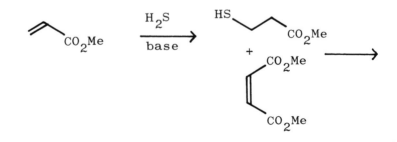

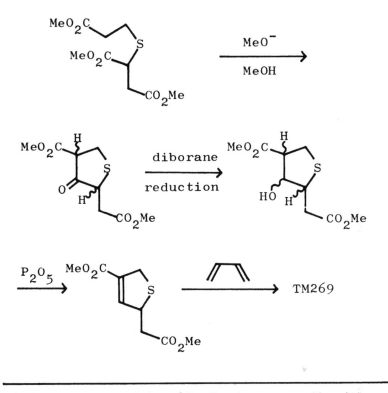

271. <u>Review Problem 27</u>: Design a synthesis
for TM271:

TM271

272. This is perhaps the most difficult problem so far; there must be many possible solutions and I can give only one.

<u>Analysis:</u> Since there is one CH$_2$ group
next to the nitrogen atom, this might be a
carbonyl group in our usual method of amine

synthesis (frame 237).

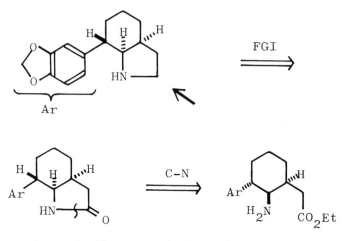

We must next disconnect the six-membered ring
and the only way we know to set up these
chiral centres specifically is by the Diels-
Alder reaction. Two alternative sites for the
double bond are possible if we convert our NH_2
to give the necessary activating group: (NO_2)

a)

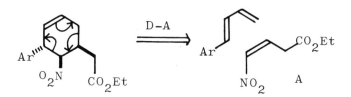

b)

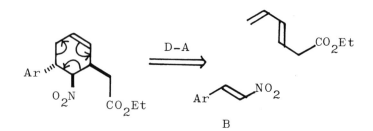

While we can make the <u>trans</u> nitro-alkene B
easily enough, the <u>cis</u>-nitro alkene A would
present problems, so let's try route <u>b</u>:

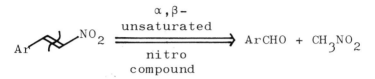

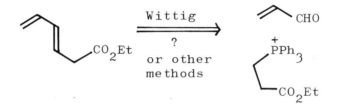

<u>Synthesis</u>: This is the method used in the
synthesis of TM271 as an intermediate for α-
lycorane (<u>J. Amer. Chem. Soc.</u>, 1962, <u>84</u>, 4951).

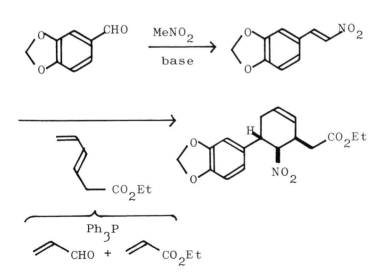

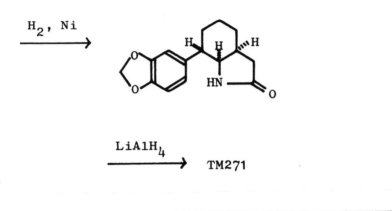

1. THREE-MEMBERED RINGS

273. Three membered rings are kinetically easy
to form but are rather unstable. Some con-
ventional methods work but are rather capri-
cious. This obvious disconnection on cyclo-
propyl ketones turns out to be all right:

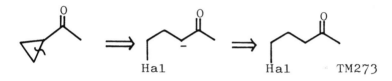

How would you make the γ-haloketone TM273?

---

274. The obvious disconnection gives an epoxide
and an enolate anion:

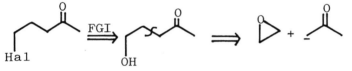

We have already explored this route (frame 184)
and found that the reaction sequence actually
is:

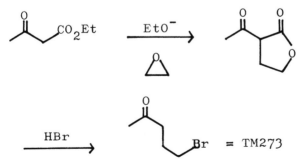

How then would you make TM274?

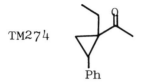

TM274

---

275. <u>Analysis</u>: We must consider two alternative disconnections of the three-membered
ring:

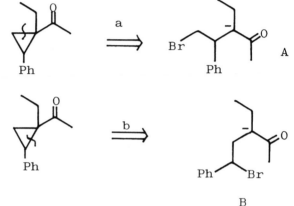

Since epoxides are attacked by anions at the
less substituted carbon atom, we shall be able
to make B but not A, so we continue.

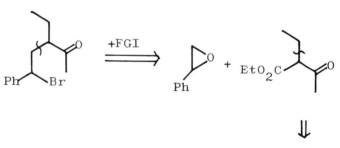

etc.

Synthesis:

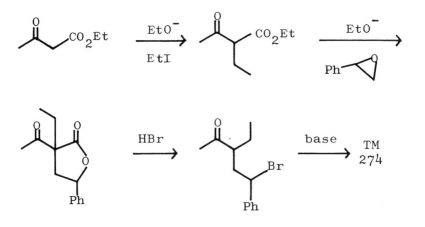

---

276. A more general route to three-membered rings is based on a new type of disconnection: the removal of an atom

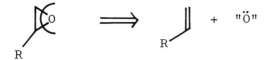

The synthon $\ddot{X}$ will be in a lower valency state - if **X** is C then it will be a carbene or the synthetic equivalent of a carbene. Let's see how this disconnection works out for epoxides.   Taking X = O first we have

$$ \begin{array}{ccc} \includegraphics[]{} & \Longrightarrow & \end{array} $$

A reagent for the synthon $\ddot{O}$ is a peracid RCO_3H.

How then could you make TM276?

TM276

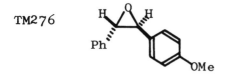

---

277. <u>Analysis:</u> Using our new disconnection:

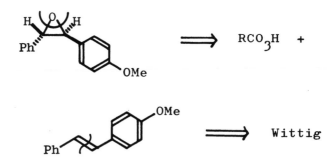

Note that is must be the <u>trans</u> olefin as it
is the <u>trans</u> epoxide we want. This is all
right as the Wittig reaction can easily be
controlled to give mostly the more stable
<u>trans</u> olefin.

   <u>Synthesis:</u>

$PhCH_2Br \xrightarrow[\text{2.base}]{\text{1.Ph}_3\text{P}} PhCH=PPh_3 \xrightarrow{\text{ArCHO}}$

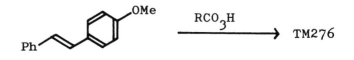

278. The reagent for the synthon $\overset{\shortparallel}{O}$ changes
when the olefin is electrophilic as in an
α,β-unsaturated carbonyl compound: then
alkaline hydrogen peroxide (HOO^-) is used.
How could you
make TM278?

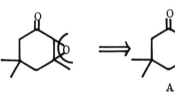

TM278

---

279. Analysis:

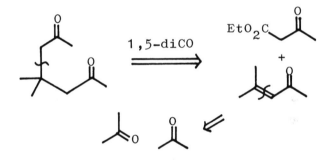

Synthesis: Intermediate A is manufactured
but can be made by the route devised here:

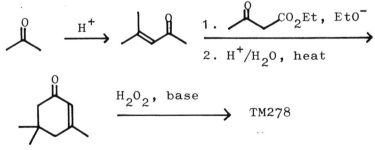

spontaneous cyclisation

---

280. The alternative disconnection for an epoxide:

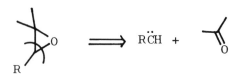

is also useful for epoxides of α,β-unsaturated carbonyl compounds when it becomes:

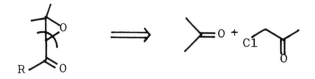

This is the Darzens reaction (frames 172-3) (see Norman, p.231 if you want more details). How would you make

TM280

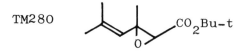

---

281. <u>Analysis</u>: Using the carbonyl group as a guide:

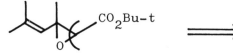

Synthesis:

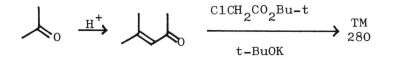

---

282. The same disconnection can be used with
simple epoxides when the sulphur ylid (A) is
used a reagent for the synthon CH_2. Draw a
mechanism for the reaction:

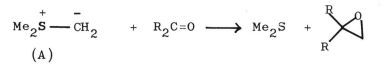

(A)

283.

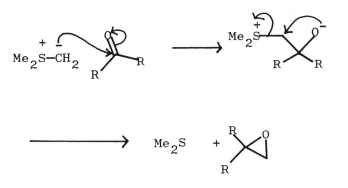

Notice that sulphur ylids behave quite dif-
ferently from phosphorus ylids, which would of
course do the Wittig reaction (frames 41-43).

How could you make TM283?

TM283

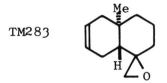

---

284. <u>Analysis</u>: Using our sulphur ylid disconnection:

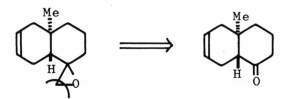

This looks like a Diels-Alder product, but the stereochemistry is wrong. However, the centre next to the C=O can be epimerised in base so:

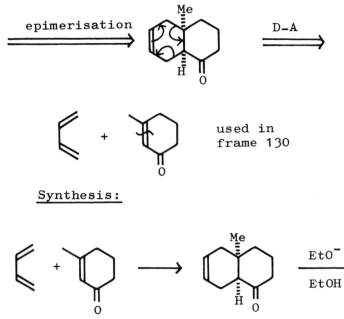

used in frame 130

<u>Synthesis:</u>

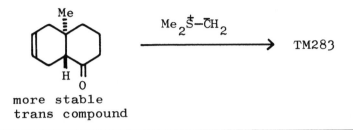

more stable
trans compound

TM283

---

285. The same disconnection is also effective
for cyclopropanes but the reagent for the
carbene synthon is a diazocompound RCHN$_2$ or a
dihalo compound treated with a metal e.g.

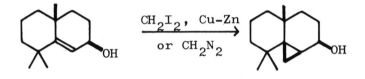

How then might you make TM285?

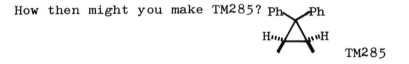

TM285

---

286. <u>Analysis:</u> Two disconnections are possible
but the stereochemistry gives us the clue:

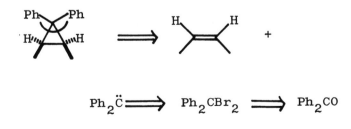

Synthesis:

$$Ph_2CO \xrightarrow{PBr_5} Ph_2CBr_2 \xrightarrow[\text{MeLi}]{} TM285$$

---

287. Diazo compounds are particularly easy to make with the diazo group next to a carbonyl group by this reaction:

$$RCOCl \xrightarrow{CH_2N_2} RCO.CHN_2 \xrightarrow[\text{or light}]{Cu(I)} RCO.\ddot{C}H$$

So how would you make TM287?

TM287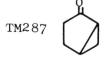

---

288. <u>Analysis:</u> Using the carbonyl group as a guide:

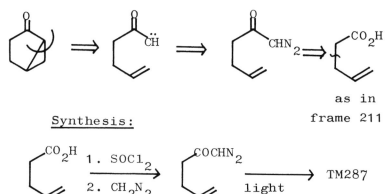

as in
frame 211

Synthesis:

TM287

Similar to <u>Tetrahedron</u>, 1964, <u>20</u>, 1807.

2. FOUR-MEMBERED RINGS

289. The most important disconnection for four-membered rings corresponds to the photochemical 2 + 2 cycloaddition of olefins:

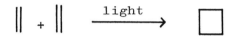

This is the allowed process by the Woodward-Hoffmann rules (see Tedder, Part 3, pp.383-387 or Norman p.292ff if you want to know more). There are obviously two disconnections for any given cyclobutane but it is often easy to see the better at once. How would you make:

TM289

---

290. <u>Analysis</u>: The two possibilities are:

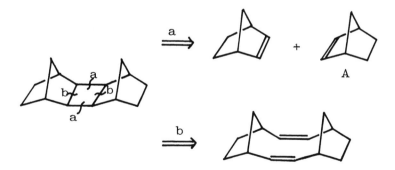

A

Far more is achieved by <u>a</u>, and TM289 is simply synthesised by irradiating norbornene (A).

The stereochemistry turns out all right. How
about TM290?

TM290

---

291. <u>Analysis:</u> TM290 was used by Corey (<u>J</u>.
<u>Amer</u>. <u>Chem</u>. <u>Soc</u>., 1964, <u>86</u>, 1652) in his
synthesis of α-caryophyllene alcohol. Again,
only one disconnection is very productive:

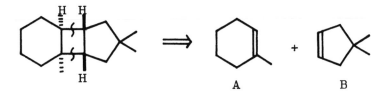

A is simply made, but the synthesis of B is
not straightforward and it turns out that it
is best done not from a mono-alcohol, but
from a diol:

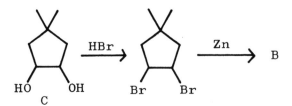

How might we make the diol (C)?

---

292. Analysis: Our methods for making 1,2-
dioxygenated compounds (frames 154-157)
involve reductive linking of a dicarbonyl
compound:

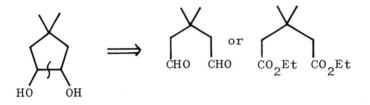

These are 1,5-dicarbonyls so we shall make
them by Michael reactions e.g.

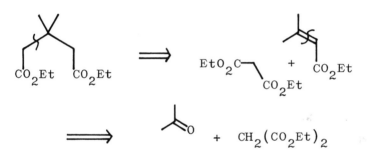

the corresponding sequence with the aldehyde
is not so easy to do. So one synthesis for
C would be:

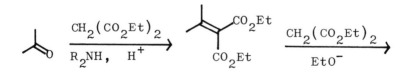

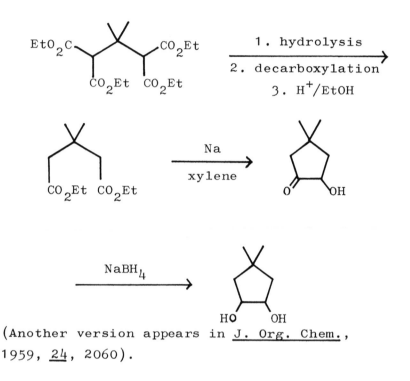

(Another version appears in <u>J. Org. Chem.</u>,
1959, <u>24</u>, 2060).

---

293. Just to show you that you can't auto-
matically separate two rings to get the right
disconnection (hint!). How could you make
TM293?

TM293

---

294. <u>Analysis:</u> Opening the cyclobut<u>ene</u> gives
this:

Now we need a Diels-Alder reaction so we must shed a double bond:

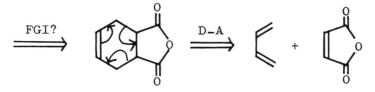

Synthesis: The FGI is easily done and the product was used by Van Tamelen (J. Amer. Chem. Soc., 1963, 85, 3297) in one of the early syntheses of a Dewar benzene:

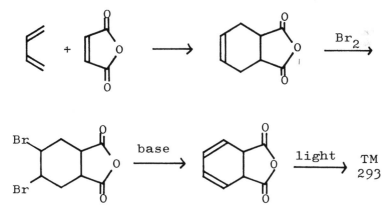

_____

3.   REVIEW PROBLEMS

295. Review Problem 28: Design a synthesis

for TM295; an intermediate needed for a tri-
chodermin synthesis.

TM295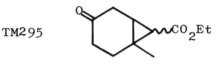

---

296. <u>Analysis:</u> Disconnecting the three-
membered ring first:

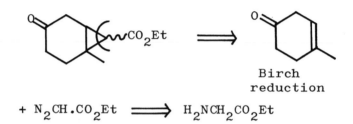

Birch
reduction

$+ N_2CH.CO_2Et \implies H_2NCH_2CO_2Et$

<u>Synthesis:</u> The route used in <u>Chem. Comm.</u>,
1971, 858:

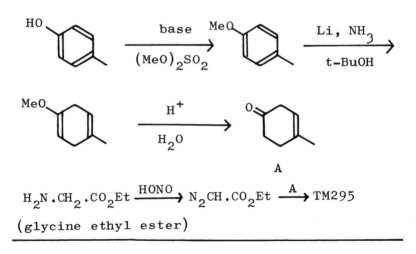

$H_2N.CH_2.CO_2Et \xrightarrow{HONO} N_2CH.CO_2Et \xrightarrow{A} TM295$
(glycine ethyl ester)

---

297. <u>Review Problem 29</u>: Suggest a synthesis of TM297.

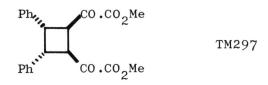

TM297

---

298. <u>Analysis</u>: Symmetry suggests that one of the two possible cyclobutane disconnections is better than the other:

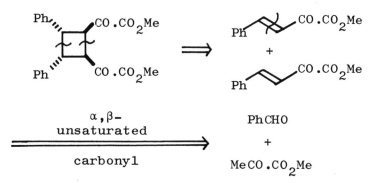

The keto-acid is the readily available pyruvic acid; there are many possible syntheses of this.

   <u>Synthesis</u>: The reactions were actually carried out like this: (Stereo- and regio-selectivity turn out all right, <u>J. Amer. Chem. Soc.</u>, 1924, <u>46</u>, 783).

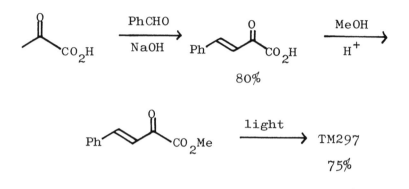

80%

light
⟶ TM297

75%

---

299. <u>Review Problem 30</u>: This may look rather difficult, but concentrate on the small ring and use the disconnection you know. Suggest a synthesis for TM299.

TM299

---

300. <u>Analysis</u>: This cage-like structure contains 6, 5, and 4-membered rings: we are most interested in the 4, and can disconnect this in two ways:

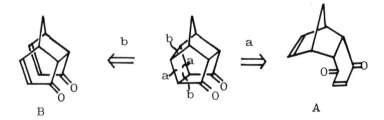

No doubt we could continue with B by disconnecting the α,β-unsaturated carbonyl groups,
but intermediate A should be recognisable as
a Diels-Alder product, and this is the
shorter route:

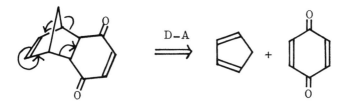

both these starting materials are readily
available.

Synthesis: Amazingly, this complicated
molecule is made in just two steps!

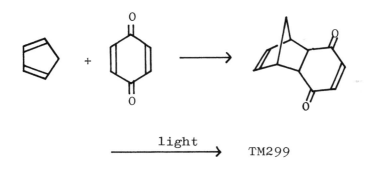

301. These problems may involve any disconnections from any section and should be useful to you for revision material.

Review Problem 31: Design a synthesis
for TM301.

$$PhCH_2O.CH_2CH_2C{\equiv}C.CH_2OH \qquad TM301$$

---

302. Analysis: A simple problem to get you
started - disconnect the unprotected side of
the acetylene first.

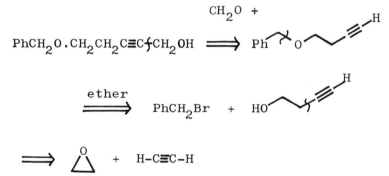

Synthesis: (As described in J. Amer.
Chem. Soc., 1966, 88, 3859).

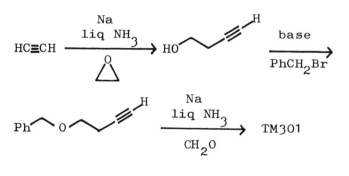

---

303. <u>Review Problem 32</u>: Design a synthesis
for TM303, an intermediate used in the synthesis of substances found in ants.

TM303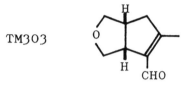

---

304. <u>Analysis</u>: We must first decide whether
to disconnect the ether ring or the α,β-
unsaturated aldehyde. No doubt reasonable
routes could be produced for both, but I
shall give one only:

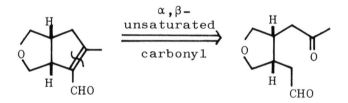

This is a 1,6-dicarbonyl compound so we must
're-connect' into a cyclohexene.

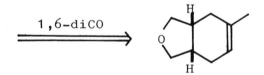

This has the oxygenation pattern of a Diels-
Alder adduct if we convert it to a carbonyl
compound.

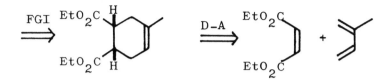

Synthesis: A successful synthesis

(J. Amer. Chem. Soc., 1958, 80, 3937) by this

route actually uses maleic anhydride:

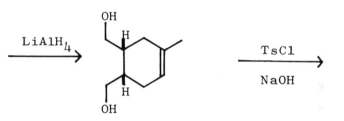

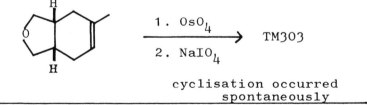

cyclisation occurred
spontaneously

305. <u>Review Problem 33</u>: Design a synthesis for
TM305, introduced in 1974 as the anti-
inflammatory drug 'clopinac'.

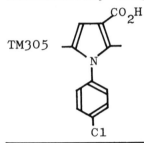

TM305

306. <u>Analysis</u>: The most sensible place to
start is with the N–C=C bonds as that will
give us some carbonyl groups (as in frames
253-7).

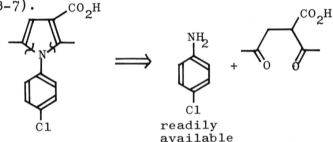

readily
available

1,4-dicarbonyl, guided by the presence of the
extra CO_2H group.

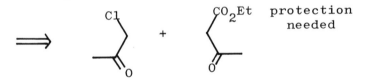

protection
needed

Synthesis: It's difficult to find out how
drug companies make their products (under-
standably!) so we can only speculate:

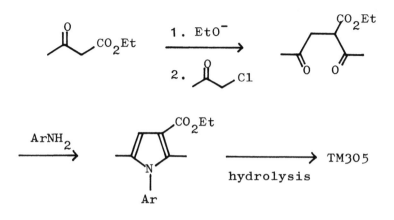

---

307. <u>Review Problem 34:</u> Design a synthesis for rose oxide, TM307, a perfume occuring in rose and geranium oils which is made at present by the oxidation of another natural product, citronellol.   TM307

---

308. <u>Analysis:</u> The cyclic ether can clearly be made from the open-chain diol:

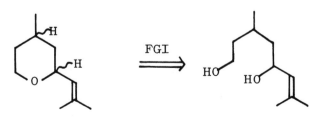

We now have a 1,5-diO relationship which
could be made by a Michael reaction if we have
two carbonyl groups.

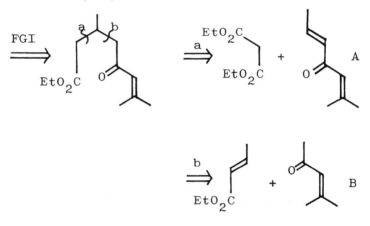

Synthesis: You will see that there are
problems in both the routes found by the
analysis. For route a it is known that
malonate attacks exclusively the less hindered
side of some Michael acceptors:

$(EtO_2C)_2CH_2$ +

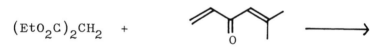

$(EtO_2C)_2CH$

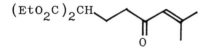

but whether it would be so selective for A
with an extra methyl group is doubtful.   For
route b , the problem is to activate the
methyl group of B and one method which might
work is:

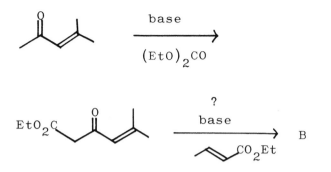

The last stages are simple enough.

The modern organic chemist has a variety both of reagents and reactions far beyond those we have looked at here. If you study organic chemistry to a more advanced level you should meet many of them but you will find that the principles of their design and use are the same as those you have learnt in this pro-gramme. We have now finished the basic types of disconnection and must look at the strategy of synthesis.

1.   CONVERGENT SYNTHESES

309. We have so far dealt mainly with the
tactics of synthesis: "what is a good disconnection?" and only occasionally with strategy:
"by what series of disconnection, even those
which do not initially look very good, can I
get back to good starting materials?"   Now we
must consider strategy - the overall plan of
the synthesis.   Our first criterion of a good
synthesis is that it must be short.   Work out
for yourself the yield in each of these two
syntheses: the yield in every step being 90%.

$$1. \quad A \longrightarrow B \longrightarrow C \longrightarrow TM$$

$$2. \quad A \longrightarrow B \longrightarrow C \longrightarrow D \longrightarrow E \longrightarrow TM$$

---

310. Yield for the 3-step synthesis (1.) is 73%
      Yield for the 5-step synthesis (2.) is 59%

By the time you get to a ten step synthesis,
the "arithmetic demon" ensures that the yield
is down to a miserly 35%. And this with 90%
yield in each step! So clearly a short synthesis is a good one. But we can cheat the
arithmetic demon by making our five steps
convergent rather than linear.   The convergent
version is this:

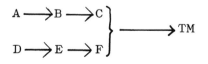

If the yield of each step is again 90%, what
is the overall yield of this five step convergent synthesis?

---

311. Clearly it is the same as the three step
synthesis, 73%. So it is better to assemble
two more or less equal parts of a molecule
separately and join them together at the end.
Let's look at this in practice. What three
alternative disconnections are immediately
obvious on TM311?

TM311

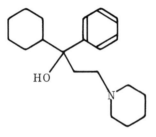

---

312. Analysis: Clearly we can remove any of
the three groups from the tertiary alcohol.

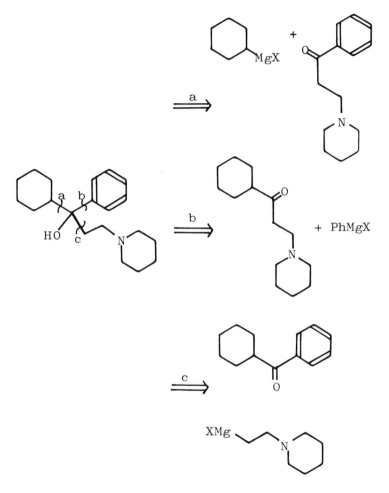

Seeing the aromatic ketones in a and c we
might have a Friedel-Crafts reaction in mind.
Continue these two a stage further.

313.

a)

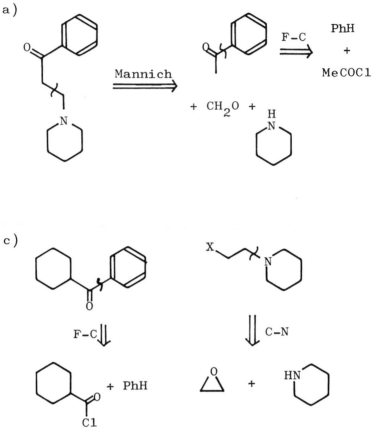

c)

Perhaps you can see that one of these routes
is linear and the other convergent.   Write
both out   in full as <u>syntheses</u> to make this
clear.

314.

a)

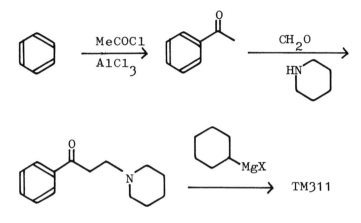

Linear

c)

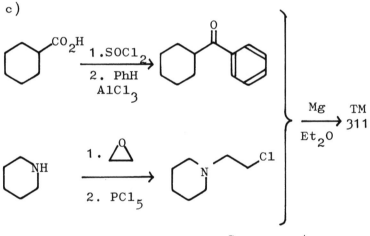

Convergent

TM311 is in fact the anti-Parkinson's disease
compound trihexylphenidyl and is made industrially by route c̲ (Tedder, volume 5, p.418).

315. You may like to reflect on our criteria for good disconnections (see for example frame 76). Two of them could be called simply guides to help us find convergent syntheses. Which ones?

___

316. 1. the greatest possible simplification.

2. the use of a branch point.

With this in mind, suggest a convergent synthesis for TM316.

TM316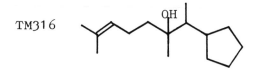

___

317. <u>Analysis</u>: Obviously we have to disconnect one of the groups next to the tertiary alcohol: two (<u>a</u> or <u>b</u>) give us plenty of simplification but only one (<u>a</u>) leads us back to a branch point:

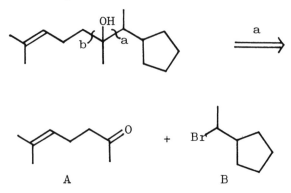

If you haven't yet considered it: how would you make A and B?

---

318. One of our standard disconnections on a ketone (frame 56) gives us a synthesis of A from TM31.

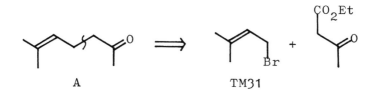

A   TM31

B can be made by the usual methods:

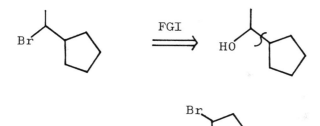

Synthesis:

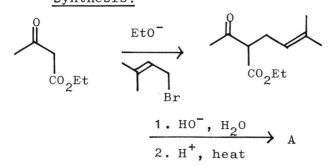

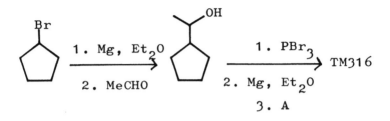

There are undoubtedly many other good
approaches; as far as I am aware, this molecule has not been made by this route.

2.   STRATEGIC DEVICES
(a)   C-HETEROATOM   BONDS

319. You have already seen that a carbon-
heteroatom bond is easy to make, since we used
such bonds as natural places for disconnections
(frames 234ff).   It is good strategy therefore
to make a carbon-heteroatom bond and then to
transform it into a carbon-carbon bond.   The
Claisen rearrangement is one way to do this:
an <u>ortho</u> allyl phenol (B) made from an allyl
ether (A):

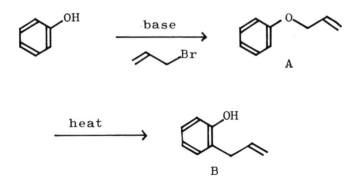

Draw a mechanism for the second step (A to B).

---

320. The reaction begins with a pericyclic step:

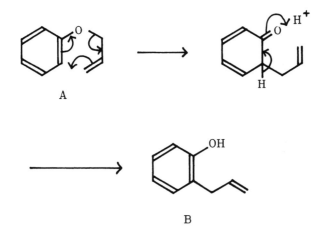

A

B

How then would you make eugenol (TM320), a constituent of oil of cloves?

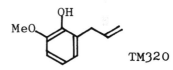

TM320

---

321. Analysis: The ortho arrangement of OH and allyl is the clue:

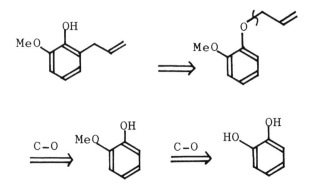

Synthesis: From readily available catechol (A):

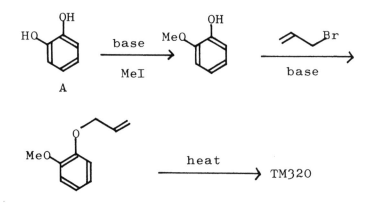

---

322. In an important industrial process, the
"Carroll reaction", an aliphatic version of
the Claisen rearrangement occurs. See if you
can find the right mechanism:

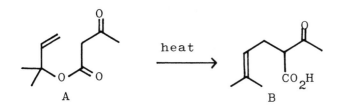

A   B

---

323. The trick is to make the enol - the
stable enol of the β-keto ester:

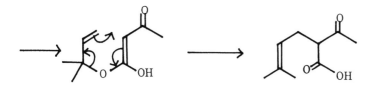

The ester 322A is made by ester exchange with
ethyl acetoacetate and a suitable alcohol.
The product 322B decarboxylates spontaneously
on heating. Draw out the whole sequence starting from ethyl acetoacetate.

324.

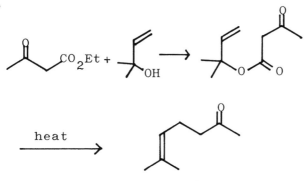

A

We made this important intermediate (A) in a
slightly different way (frame 318), but this
is how it's made industrially for use in
perfumes and flavours (Pure Appl. Chem., 1975,
43, 527).   How would you extend this synthesis
to make TM324?

TM324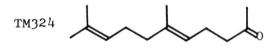

---

325. Analysis: Reversing the Carroll reaction:

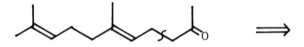   ⟹

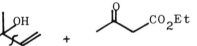   +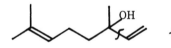

⟱

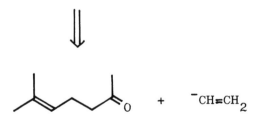   +   ⁻CH=CH₂

The vinyl anion synthon is best represented by an acetylide ion (frame 33).

Synthesis:

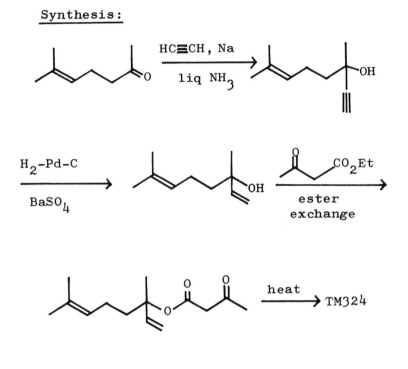

326. Another similar example concerns the alkylation of enamines. This reaction works well with reactive α-halocarbonyl compounds (frames 175ff) but simple alkyl halides often react on nitrogen:

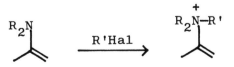

Allyl halides do however give us good yields
of alkylation at carbon:

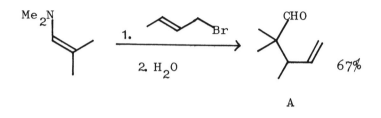

A

Suggest how this might happen.

---

327. Since the allyl group has rearranged, it
may have added to nitrogen first:

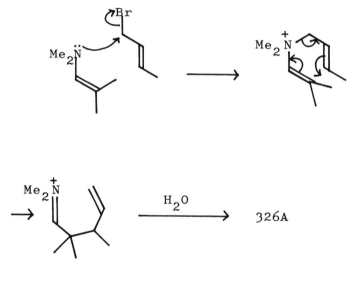

(J. Org. Chem., 1961, 26, 3576).

---

328. The strategy of using intramolecular
reactions to set up the correct relationship

between two groups is of more general importance.   We obviously want to disconnect bonds
a and b in TM328
so that we add
a four carbon
fragment to
PhOMe in the   TM328
synthesis.   It
will be easy to

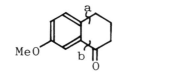

put in bond a because it is para to the MeO
group, but bond b might be difficult.   The
strategy is to put in bond a first and use an
intramolecular reaction to force bond b to go
in the right place.   Succinic anhydride is a
convenient four carbon electrophile:

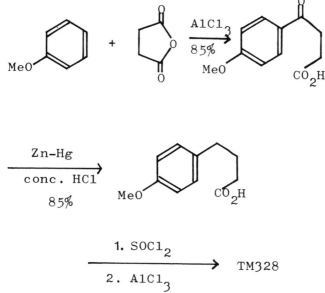

This approach is described in <u>Org. Synth.</u>
<u>Coll.</u>, <u>2</u>, 81, 499, 571, and these particular
compounds and reactions in <u>J. Chem. Soc.</u>,
1934, 1950; <u>J. Amer. Chem. Soc.</u>, 1936, <u>58</u>,
1438, 2314; 1942, <u>64</u>, 928. The final cyclisation can be done directly on the acid with
anhydrous HF in 93% yield.

(b)   <u>POLYCYCLIC COMPOUNDS -</u>
      <u>THE COMMON ATOM APPROACH</u>

329. Another strategic device applies specifically to polycyclic compounds. In the inte-
rests of simplification we want to remove some
of the rings and give an intermediate with a
familiar ring structure. We can do this by
the "common atom" approach. In TM329, mark
all the carbon atoms which belong to more than
one ring - the "common atoms".

TM329

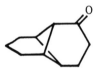

330.

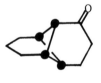

● = common atoms

Now disconnect any bond joining two common
atoms and see if there is a good starting
material.

---

331. Because of the symmetry of TM329 there
are only two different disconnections of bonds
between two common atoms.

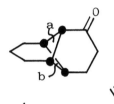

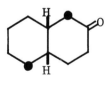 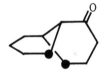

|| ||

 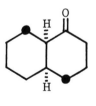

A   B

One of the marked atoms in each intermediate
will have to be + and one -. Can you see a
good starting material here?

---

332. For A we can put the - next to the carbonyl group and provide a leaving group for
the + :

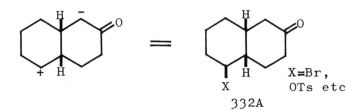

332A   X=Br,
                                        OTs etc

If you don't see why the stereochemistry
should be as I have drawn it, I suggest you
make a model of 332A and discover for yourself.
There is a simple synthesis of 332A (X = OTs)
from the Robinson annelation (frame 117)
product 332B.

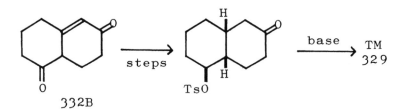

332B

Using the common atom approach, design a synthesis of TM332.

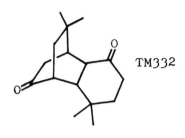

TM332

333. <u>Analysis:</u> Marking the common atoms we find there are three possible disconnections of bonds between them, but only <u>a</u> or <u>c</u> give us simpler precursors. Both also are 'logical' in that we can immediately write reagents for the synthons.

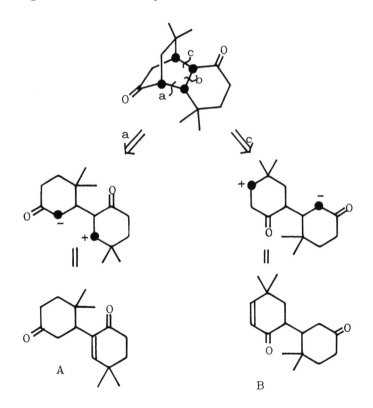

Both A and B are 1,5-dicarbonyl compounds, but
only B can be disconnected in the usual way.
The result is two molecules of an α,β-unsatu-
rated ketone and we can continue the analysis:

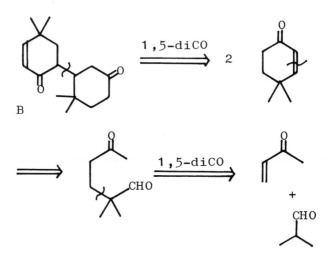

Synthesis: Carried out by Lewis (J. Chem.
Soc. (C), 1971, 753) using enamines for each
step:

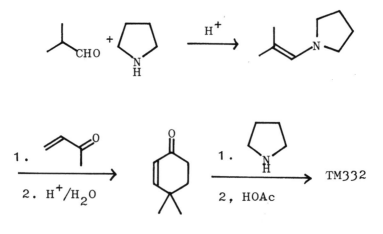

3.   CONSIDERING ALL POSSIBLE DISCONNECTIONS

334. Apart from aiming for a convergent synthesis or using the two strategic devices we've
just seen, special tricks in strategic plan-
ning are unimportant. The main thing is to
find the shortest route with the best individ-
ual steps. To do this we ought to consider all
possible disconnections even those which don't
look very promising initially.

As an example, let's analyse the synthesis of   γ-lactones (e.g. TM334) and see how
we may choose one of a number of strategies
depending on the structure of the target molecule. We'll consider in turn each of the
three C-C bond disconnections. The one with
the most appeal is probably b: complete the
analysis for this approach.

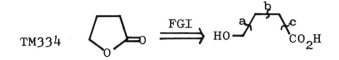

TM334

---

335. Analysis: This is the approach to 1,4-
dioxygenated skeletons we used in frames 171-
186. We need an 'illogical' electrophile - in
this case an epoxide:

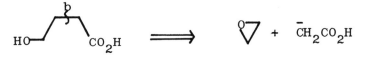

<u>Synthesis</u>: Protection and activation as usual:

$CH_2(CO_2Et)_2$ $\xrightarrow{\text{EtO}^-,\ \triangle}$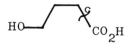

$\xrightarrow[\text{2.H}^+,\ \text{heat}]{\text{1.HO}^-/\text{H}_2\text{O}}$ TM334

---

336. What possible synthons might we need for disconnection <u>c</u> :

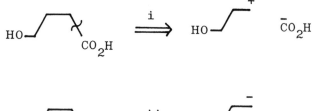

---

337. We can choose either polarity:

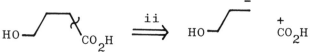

Both are possible but we are more used to (<u>i</u>) since we can use a Michael acceptor and cyanide ion for the two synthons. How would you actually do a synthesis this way?

338.

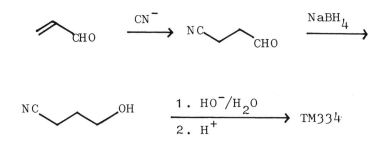

---

339. Now analyse the last disconnection, <u>a</u>, in the same way, writing the synthons and considering possible reagents for them.

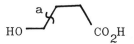

---

340. Again, either polarity is possible:

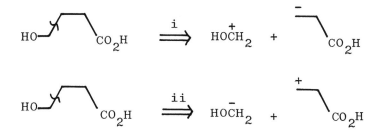

Route (<u>ii</u>) could use a Michael acceptor together with a suitable one carbon nucleophile such as MeOCH$_2$MgCl. Route (<u>i</u>) could have

formaldehyde as the electrophile and a suit-
ably protected and activated derivative for
the nucleophile such as . . .

---

341. By analogy with the Reformatsky reaction,
the zinc derivative of a β-bromoester would do

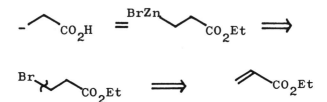

If this chemistry is successful, the synthesis
becomes:

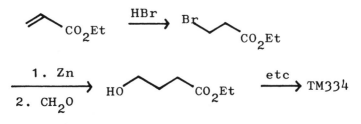

---

342. Each of these approaches may be the best
for any given lactone: the one in the last
frame for example would allow you to use any
Michael acceptor and any aldehyde.

Which strategy is being followed here?

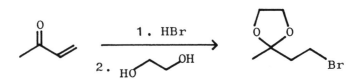

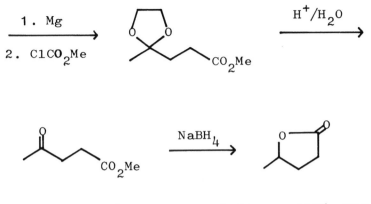

(<u>Tetrahedron Letters</u>, 1976, 3105)

---

343. One we hadn't explored in detail: it's
strategy <u>c</u>(<u>ii</u>) outlined in frame 337 with
$ClCO_2Me$ for $^+CO_2H$ and the protected β-bromo-
ketone for $MeCH(OH)CH_2CH_2^-$.

---

344. Now try this one for yourself: Suggest
which strategy is most suitable for TM344.

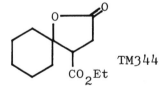

TM344

---

345. <u>Analysis:</u> Opening up the ring to see the
true problem & using our branch point guide we
really want to disconnect the bond between the
two ringed atoms so that strategy <u>a</u>(<u>i</u>), frame

340 is best.

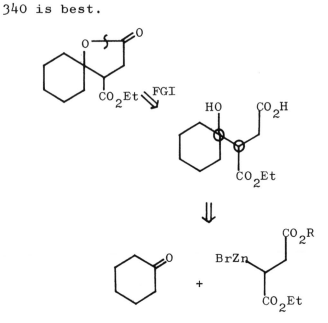

Synthesis: Putting R = Et, this is easy:

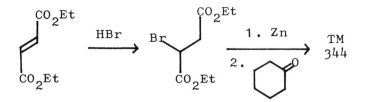

346. Now try your hand at this more challenging example.

TM346

347. Here is one possible solution:

Analysis: As usual, we open the ring first: the tertiary alcohol can be made by adding one or two mols of MeMgI.

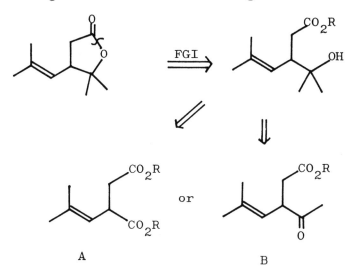

A is no good because MeMgI would have to attack the more crowded ester first. So, how do we make B?

348. Analysis: The most suitable disconnection follows the strategy we originally used (b, in frames 335, and 171-175). But can we make the right enamine from the unsaturated ketone, and does it alkylate in the right place? It turns out that it can and does.

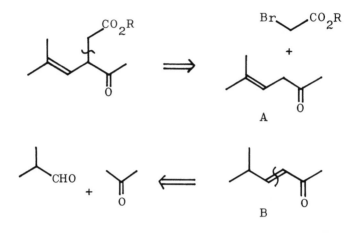

Either 348A or B gives the right enamine, but
B is an α,β-unsaturated ketone and so easier
to disconnect.

   Synthesis: (Bull. Soc. Chim. France, 1955,
1311; 1962, 2243).

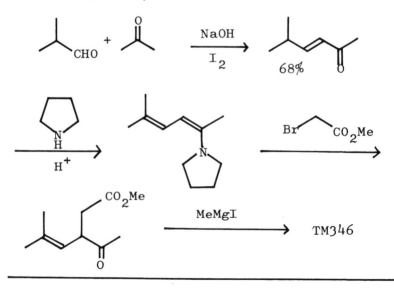

4.   ALTERNATIVE FGIs BEFORE DISCONNECTION
The Cost of a Synthesis

349. The   γ-lactone problem is made easier
because the FGs are all based on oxygen.   The
molecule can therefore be disconnected without
FGI except for oxidation or reduction.   Let's
now look at the synthesis of a molecule with a
'difficult' FG: the muscle relaxant 'baclofen'
TM349.   What is the difficult FG?

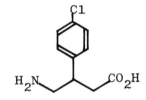

TM349

---

350. The amino group.
What FGI's on TM349 give us molecules we can
disconnect?.   There are in fact three, but you
may not see one of them.

---

351. We have meet NO_2 and CN before, but a
primary amine can also be made from an amide
by the Hofmann degradation (see Norman p.446-
7 or Tedder, vol.2, p.281-2).

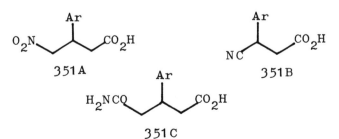

351A

351B

351C

To make 351C we need to convert a dicarboxylic
acid into its half amide. How might we do
this?

---

352. Since the molecule is symmetrical, attack
of ammonia on the cyclic anhydride will do it:

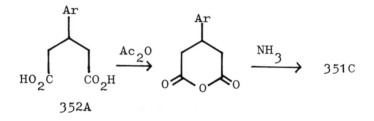

So we really need to disconnect 351A and B and
352A. Analyse possible disconnections of
these three on the same chart since they will
have several starting materials in common.

---

353. <u>Analysis:</u> All by familiar disconnections

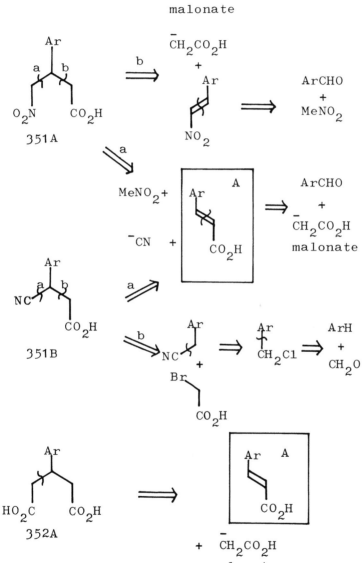

All these routes use known reactions and are
about the same length. Do you notice that no
less than _three_ have the unsaturated acid 353A
as an intermediate. If we need to try out new
reactions it is best to choose a route with a
common intermediate (353A here) so that if one
route fails we can use the same intermediate
for another. We can then choose between the
three routes on cost. The 1983 prices of the
starting materials are:

p-chlorobenzaldehyde   £9.00/500 g
diethyl malonate   £4.90/500 g
sodium cyanide   £3.50/500 g
nitromethane   £8.30/kg

Assume solvents etc., cost the same for each
route. Which route will you try first?

---

354. Cost per mole is the relevant figure:

p-chlorobenzaldehyde, MW 140.5,
        costs £2.51 per mole.
diethyl malonate, MW 160,
        costs £0.93 per mole.
sodium cyanide, MW 49,
        costs £0.34 per mole.
nitromethane, MW 61,
        costs £0.51 per mole.

353A is common to all the routes we are considering but it is obviously cheaper to use a
mole of cyanide or nitromethane rather than
another mole of malonate. In fact, though,
these contribute relatively little to the cost,
the main part being p-chlorobenzaldehyde. So,

use whichever route you like!

---

5.   FEATURES WHICH DOMINATE STRATEGY

355. Chrysanthemic acid (TM355) is an important
constituent of pyrethrins - naturally occurring
insecticides which are virtually harmless to
mammals.   What feature of this molecule will
dominate our strategic thinking?

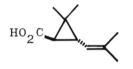

TM355

---

356. You might reasonably have said stereochemistry, but the best answer is the three
membered ring.   Which three 'carbene' disconnections might we consider?   (See frames 276-
288 if you've forgotten all this!).

---

357. We disconnect two bonds at once to give
us a 'carbene' and an olefin:

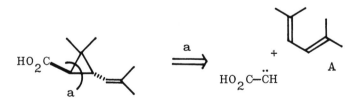

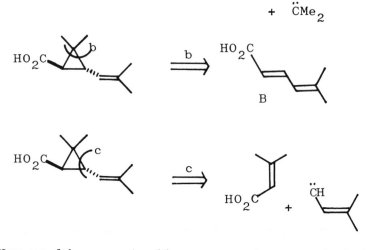

How would you actually carry out $\underline{a}$ and what
do you think of the prospects?

358. <u>Analysis</u>: The carbene synthon is easy: it
can be ethyl diazoacetate N_2CHCO_2Et.   The diene
can be made by the Wittig reaction from a
familiar allylic bromide (TM31).

   <u>Synthesis</u>:

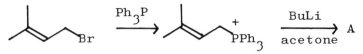

$$NH_2CH_2CO_2Et \xrightarrow{\text{HNO}_2} N_2CHCO_2Et \xrightarrow{A} TM355?$$

   <u>Comments</u>: The diene A is symmetrical so
it doesn't matter which double bond is attacked
by the carbene.   On the other hand, it may be
difficult to stop carbene addition to the

second double bond. The only control over the
stereochemistry will be that the _trans_ compound we want is more stable. Japanese
chemists have recently synthesised optically
active _trans_ chrysanthemic acid by this route
(_Tetrahedron Letters_, 1977, 2599).

It turns out that the less stable _cis_ can
be converted into the _trans_ by treating the
ethyl ester with EtO⁻ in EtOH. (_ibid._, 1976,
2441).

---

359. Now consider strategy _b_. How would you
make the diene acid B, what reagent would you
use for the carbene synthon, and how do you
rate the chances of this route?

---

360. _Analysis_: The normal approach to the
diene looks all right:

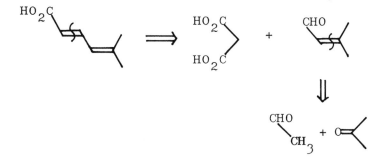

The carbene synthon might be difficult, but
since the olefin is conjugated with a carbonyl group we could try a sulphur ylid as a
nucleophilic carbene equivalent (as in frame
283).

<u>Synthesis:</u> The diene could be made by this route:

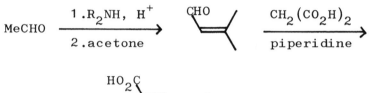

MeCHO $\xrightarrow[\text{2.acetone}]{\text{1.R}_2\text{NH, H}^+}$ [CHO] $\xrightarrow[\text{piperidine}]{\text{CH}_2(\text{CO}_2\text{H})_2}$

B

One example of a suitable sulphur ylid is the one below (C) (Corey, <u>Tetrahedron Letters</u>, 1967, 2325) and ylids of this sort have been added to α,β-unsaturated ketones (<u>Tetrahedron Letters</u>, 1966, 3681):

$\text{Ph}_2\text{S} \xrightarrow{\text{EtI}} \text{Ph}_2\overset{+}{\text{S}}\diagup \xrightarrow[\text{MeI}]{\text{base}} \text{Ph}_2\overset{+}{\text{S}}\diagleftarrow \xrightarrow{\text{base}}$

$\text{Ph}_2\text{S}=\diagleftarrow$ C $\xrightarrow[\text{(ester)}]{\text{B}}$ TM355? (as the ester)

<u>Comments:</u> We have to explore new chemistry here and the main problem does seem to be getting the right combination of diene and carbene reagent. Perhaps not such a good route.

---

361. However, things don't always conspire against us in synthesis! In 1976 Krief (<u>Tetrahedron Letters</u>, 1976, 3511) carried out an innocent-looking Wittig reaction which might reasonably have given the ester of diene 360B.

Instead it gave the ester of TM355 in good
yield!   Can you explain what has happened?

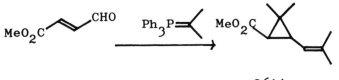

361A

---

362. First the normal Wittig reaction, and
then a second mol of Wittig reagent must be
behaving as a carbene equivalent, just as we
hoped the sulphur ylid would (but
see Tetrahedron Lett., 1979, 1511, 1515).

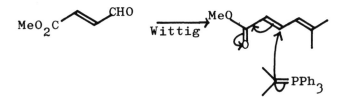

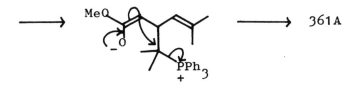   361A

So this strategy turns out to be all right too.
Now how might you realise strategy c in frame
357?

---

363. Analysis: The unsaturated acid (or better
ester) is easy enough while the carbene reagent might be a sulphur or phosphorus ylid
based on our old friend TM31:

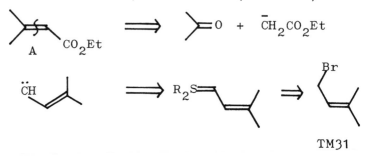

Synthesis: Though the ester 363A has been
made this way (J. Indian Chem. Soc., 1924, 1,
298) in 60% yield, the rest of the synthesis
has not yet been tried as far as I know.

364. Perhaps the most sensational synthesis of
chrysanthemic acid uses this strategy. You
may remember that TM31 is usually made from
the adduct of acetylene and acetone. Draw out

the stages of this reaction sequence.

365.

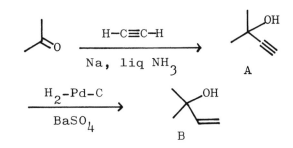

Raphael has devised a commercial synthesis
using intermediates 365A and B to provide the
two halves of the molecule.   B is converted in
aqueous acid to an isomeric alcohol.   Draw
this.

366.

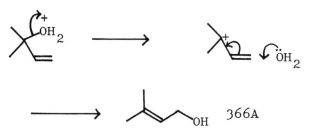

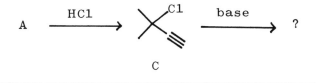

366A

Now A is converted into its chloride, (C) and
this is treated with base.   Which proton will
be removed?

367. The acetylenic proton!   The carbanion now eliminates Cl⁻ to give a most odd-looking carbene.   Can you see what it is?

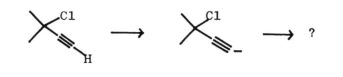

368.

This allenic carbene is now added to the alcohol 366A.   What will be the product?

369.

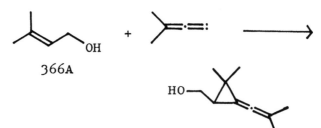

366A

Note that no stereochemistry has been introduced so far. Reduction (sodium and liquid ammonia) selectively gives trans chrysanthemic alcohol which can be oxidised to the acid with CrO_3. Draw out the whole synthesis as a chart.

370. This is Raphael's complete synthesis of
chrysanthemic acid (<u>Chem. Comm.</u>, 1971, 555;
<u>J.C.S. Perkin I</u>, 1973, 133):

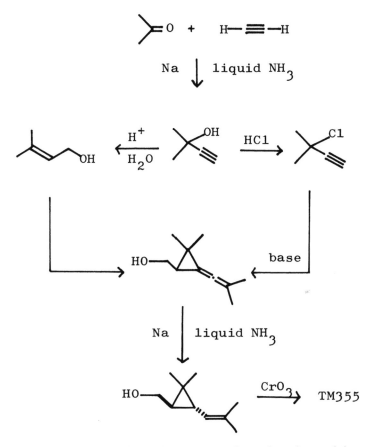

In all, only six steps are involved, making
this a most economical synthesis. Chrysanthemic acid is important enough to have been
made in many other ways too (e.g. <u>Tetrahedron</u>
<u>Letters</u>, 1976, 2441; <u>Bull. Soc. Chim. France,</u>
1966, 3499).

6. FUNCTIONAL GROUP ADDITION
(a) STRATEGY OF SATURATED HYDROCARBON SYNTHESIS.

371. Perhaps you can see from these examples that it is possible to mould quite poor-looking initial disconnections into good over-all strategies. We need to be particularly flexible in designing the synthesis of hydrocarbons since we have no functional groups to guide us. Take 'twistane' (TM371) for example. This molecule has a six-membered ring (numbered) fixed in the twist boat conformation and the molecule was needed to study this unusual feature. In designing a synthesis of twistane, which bonds might we want to disconnect?

TM371
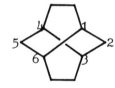

---

372. Using our strategic device of marking atoms common to more than one ring (see frames 329-333), we find there is only one disconnection of a bond joining two common atoms as the molecule is symmetrical:

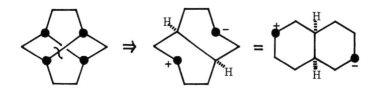

Now we must introduce some functional groups
to turn this synthon into a reagent.   What do
you suggest?

___

373.   The obvious ones are a carbonyl group
next to the anion and an OH group (which can
easily be converted into a leaving group) for
the cation giving two possibilities:

One of these has a familiar oxygenation pattern.
Which?

___

374. A has the same oxygenation pattern as the
Robinson annelation product 374A (frames 117-8).

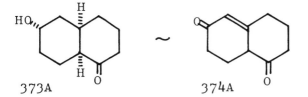

How might the conversion be done?

---

375. The saturated ketone is more reactive
than the conjugated ketone so we shall need to
protect the one before reducing the other.

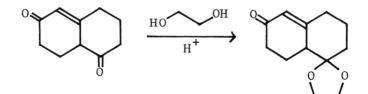

$$\dfrac{1.H_2, \text{ catalyst}}{2.H^+/H_2O} \longrightarrow \quad 373A$$

Now draw out the rest of the synthesis, before
and after 373A.

376.

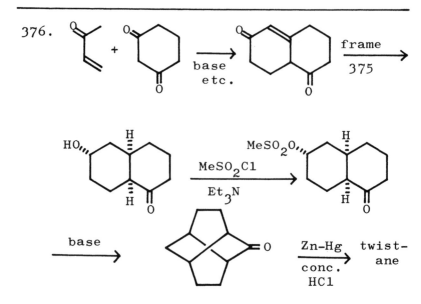

Twistane has been synthesised using this
approach, though with different details, and
also by routes using three other reasonable
disconnections (b, c, and d). These are
described in the kind of language used in this
programme by Hanson and Young (Austral. J.
Chem., 1976, 29, 145).

---

377. The most important point in analysing the
synthesis of a hydrocarbon is where to put the
carbonyl group, and this can depend on features
other than common atoms. What strategies might
you use in the synthesis of TM377?

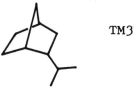

TM377

---

378. The isopropyl group could be derived
from a carbonyl group and a Grignard reagent in
two ways:

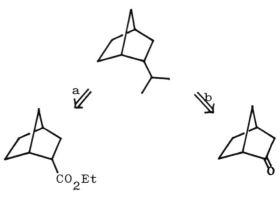

+ MeMgI   + i-PrMgBr

followed in each case by elimination of water
and reduction of the olefin.   Further thoughts?

---

379. Strategy <u>a</u> can lead back to a Diels-Alder
reaction if we put in another double bond - a
trivial step since we are going to reduce out
all double bonds at the end:

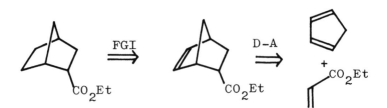

Write out the synthesis.

---

380. <u>Synthesis:</u>

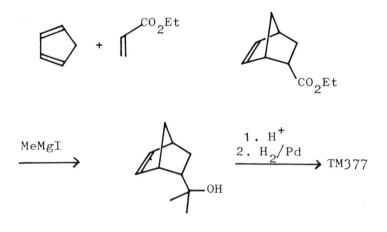

<u>(b)</u>   <u>FUNCTIONAL GROUP ADDITION</u>
       <u>TO INTERMEDIATES</u>

381.   Consider this problem: We have always
assumed that intramolecular carbonyl condensations giving 5 or 6 membered rings are pre-
ferred over those giving 4 membered rings.
But what about 7 membered rings? Some chemists
recently wanted to investigate this point and
chose to cyclise TM381 to see if A or B were
formed.   First, TM381 has to be made. How do
you suggest they do it?

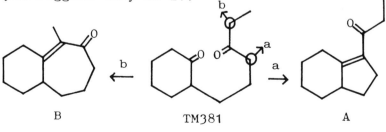

382. <u>Analysis:</u> This is a 1,6-dicarbonyl compound so a reconnection is called for.   The
next obvious series (frames 36-8) of disconnections ends up at 382B - not an easy compound to make.   Where could we put a carbonyl
group in 382A to allow some more helpful disconnections?

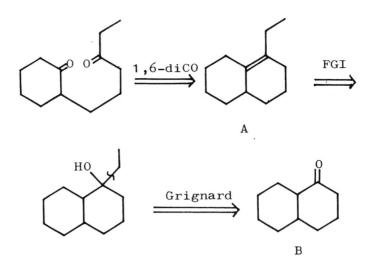

A

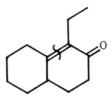

B

---

383. <u>Analysis</u> continued: I suggest to give
383A since one can then disconnect the α,β -
unsaturated ketone and get a starting material
with only one ring:

383A

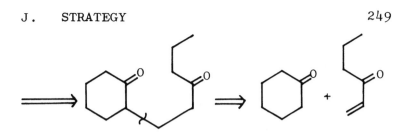

Synthesis: This was in fact done by a minor variation on our strategy, the ethyl group being added later:

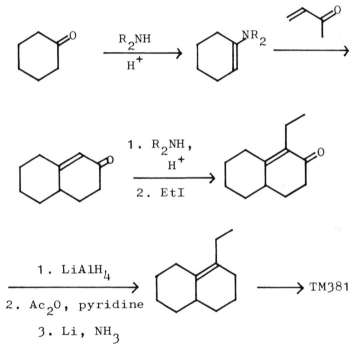

When they treated TM381 with base (MeO⁻ in MeOH) it was in fact 381A which was formed, so five membered rings are better than seven membered rings (J. Org. Chem., 1976, 41, 2955).

Though you can in principle add a carbonyl group anywhere in a target molecule, remember it means extra steps in the synthesis so use it only as a last resort.

---

7.   MOLECULES WITH UNRELATED FUNCTIONAL GROUPS

384. Saturated hydrocarbons were a problem because they have no functionality. It can be just as bad when a molecule has several functional groups all apparently unrelated. Bisabolene (TM384) has three double bonds, all rather widely separated. Comment on possible strategies in terms of the likely origin of each double bond and the probable order of events.

TM384

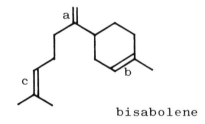

bisabolene

385.

(a)   This double bond almost certainly comes from a Wittig reaction with $Ph_3P=CH_2$ on the corresponding ketone. A good place to start.

(b)   This could come from MeMgI addition to a ketone, from a Diels-Alder reaction, Birch reduction, or Robinson annelation. Not a good place to start.

(c) This could come from a Wittig reaction
(either way round) or by addition of $Me_2C=$
$CH.CH_2Br$ (TM31) to some other compound. Quite
a good place to start.

On balance it looks best to start with a
Do the first disconnection and suggest what
might come next.

386.

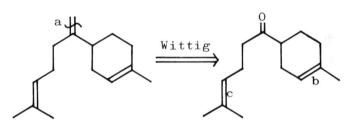

Now we can choose between the alternatives in
frame 385 for b and c . Consider these if
you've not already done so.

387. For b   the Diels-Alder now looks best.
     for c   alkylation with the allyl halide
     looks good.
There are of course other solutions, but continue the analysis along these lines.

388. If you work through both orders of events,
it turns out better to do c first and b
next:

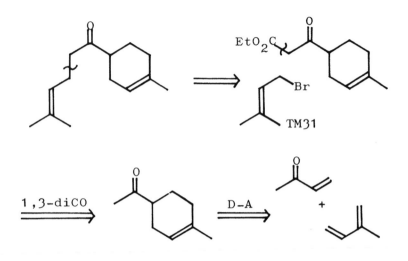

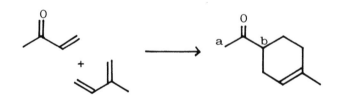

Now draw out the synthesis and comment on the selectivity we need and whether we are likely to find it.

389.

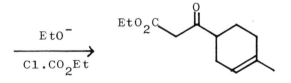

Could have given the other regio-isomer, but actually gives mostly the compound we want.

Atoms a or b could enolise but only this product can form a stable anion (frame 101) so
this is all right too.

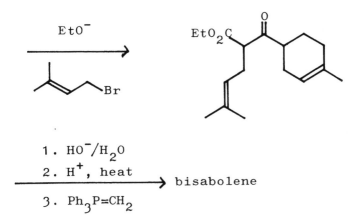

1. HO⁻/H₂O
2. H⁺, heat   ⟶ bisabolene
3. Ph₃P=CH₂

This is the synthesis carried out by Vig and
his group.   (J. Indian Chem. Soc., 1966, 43,
27).   Bisabolene has been made in several
other ways.

_____

390. You have now reached the end of the systematic instruction provided by the programme.
If you have worked your way through to this
stage and have understood most of what you
have done then you are certainly ready to move
on to more difficult problems.   The problems I
have given you so far have all been chosen
because they could be solved reasonably easily.
Many molecules present a greater challenge and,
while you will of course use the logic of the
programme in designing syntheses for them, you

will have to extend it. In the next frame you
will find some suggestions for further study
on these lines.

391. Choose some, or all of these for your
next step.

1.   Review Problems: Scattered throughout the
programme were review problems so that you
could check on your progress. These are also
useful now so that you can check that you can
still remember material you met earlier on:

Review problems 1-3   frames 29-35
Review problems 4-6   frames 78-83
Review problems 7-8   frames 108-111
Review problems 9-11   frames 125-130
Review problems 12-13   frames 167-170
Review problems 14-15   frames 190-193
Review Section:
  Synthesis of Lactones:
Review problems 16-18   frames 203-209
General review
  problems 19-23   frames 210-219
Review problem   24   frames 232-233
Review problems 25-27   frames 267-272
Review problems 28-30   frames 295-300
General review
  problems 31-34   frames 301-308

2.   Revision Problems: For those of you who
have already done all the review problems,
there are some not too difficult revision problems in the next section. All have answers.

3.   Problems in Strategy: Our discussion on
strategy was limited to the more straightforwad
aspects. This section has some challenging

problems without worked answers. These are more difficult than the revision problems. (frames 412-420)

4.   <u>Problems with several published solutions:</u> This section gives you some 'real' problems which have already been solved by several different routes. You can check your answer against these published routes. (frames 421-424)

5.   <u>Further reading:</u> Now that you know what the problems are, you will probably get a lot of value out of studying published syntheses of large molecules. Some excellent examples appear in Fleming (see book list at the front).

E. J. Corey was the originator of this analytical approach to synthesis and you might like to read some of the articles in which he first explains it. Here is a selection: <u>J. Amer. Chem. Soc.</u>, 1964, <u>84</u>, 478; 1972, <u>94</u>, 440; 1974, <u>96</u>, 6516; 1975, <u>97</u>, 6116; 1976, <u>98</u>, 189; <u>Pure Appl. Chem.</u>, 1967, <u>14</u>, 19, and <u>Quart. Revs</u>, 1971, <u>25</u>, 455.

392. <u>Revision Problem 1</u>: Leaf alcohol (TM392)
is widespread in plants and has the character-
istic smell of green leaves and grass. The
cis isomer alone has this smell and is used in
perfumery. How would you make it?

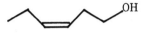

TM392

---

393. <u>Analysis</u>: The <u>cis</u> olefin can come from an
acetylene and so we are guided into our disconnections:

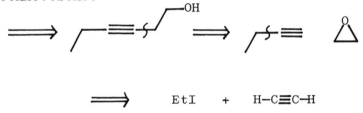

$\Longrightarrow$ EtI   +   H–C≡C–H

<u>Synthesis</u>: Sondheimer (<u>J. Chem. Soc.</u>,
1950, 877) made leaf alcohol this way.

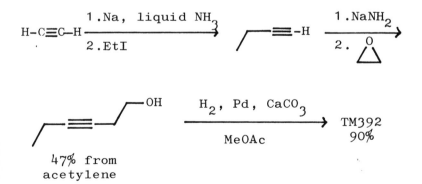

394. <u>Revision Problem 2:</u> α-Terpineol also
occurs widely in plants and was one of the
first natural products to be isolated pure.
There was originally some doubt as to whether
its structure was TM394A or TM394B. Suggest
syntheses of both these compounds so that they
can be compared with the natural product.

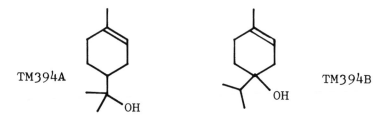

TM394A   TM394B

---

395. <u>Analysis:</u> The strategy for any modern syntheses of these compounds would be based on the
Diels-Alder reaction or the Birch reduction:

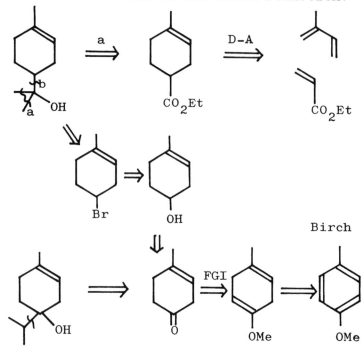

Synthesis: Since we can make both compounds from the same intermediate 395A, we'll
use the Birch reduction route:

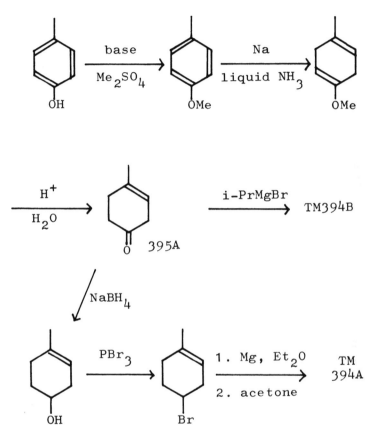

In the event, TM394A proved to be α-terpineol
and the shortest synthesis is by Alder and Vogt
(Annalen, 1940, 564, 109) using the Diels-
Alder reaction:

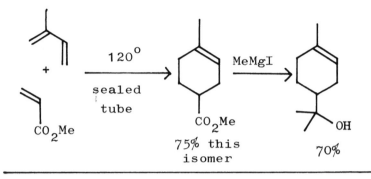

396. Revision Problem 3: House (J. Org. Chem.,
1965, 30, 1061) wanted to study intramolecular
Diels-Alder reactions and wanted molecules like
TM396 in which n is 3 or 4, so that the product will have a 5 or 6 membered ring if the
reaction works. It would obviously be a good
thing if the synthesis can easily be modified
to make other size rings as well. What do you
suggest?

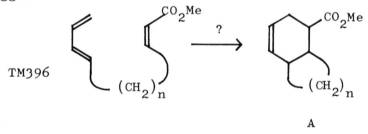

TM396

A

397. Analysis: If we take out the central por-
tion of the molecule we can use any size of n
we want. The most obvious method is two suc-
cessive Wittigs:

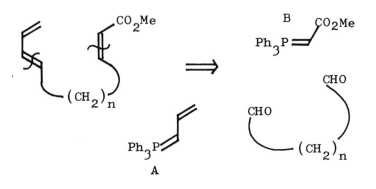

If we use the rather unreactive B first we should be able to react one aldehyde at a time. The various dialdehydes are available or can be made by the usual 1,n-dicarbonyl routes.

Synthesis: This is what House did:

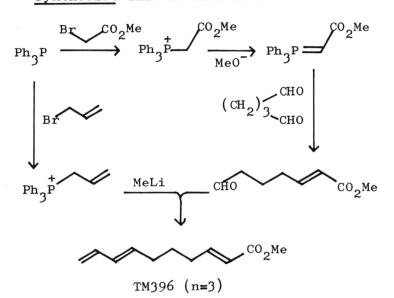

TM396 (n=3)

To complete the story, when this molecule was
heated an intramolecular Diels-Alder reaction
did indeed take place to give a new five-
membered ring, (396A, n=3).

---

398. <u>Revision Problem 4</u>: Musks are compounds
which have some pleasant smell themselves, but
function chiefly by retaining and enhancing
the perfume of other compounds. How might
'celestolide', a modern musk, (TM398) be made?

TM398
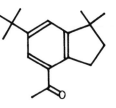

---

399. <u>Analysis:</u> The one functional group is
something of a red herring since we shall put
in the acetyl side chain by a Friedel-Crafts
reaction on the real target molecule, 399A:

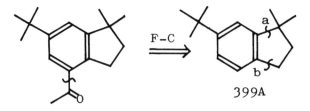

F-C

399A

This will clearly be made somehow by disconnections <u>a</u> and <u>b</u> but the order of events is
important. We must disconnect first, that is
synthesise last, the bond with the 'wrong'
orientation - i.e. <u>a</u> , <u>meta</u> to the t-butyl

group.   The reaction will then be intramolecular and orientation doesn't matter.   This
gives us 399B, and I show one possible route
from that.

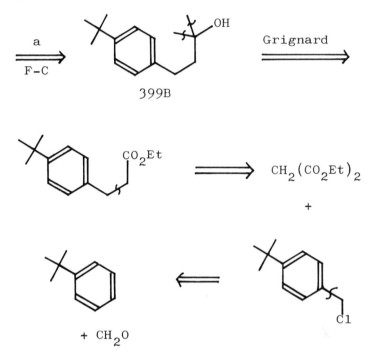

Synthesis: This route has been carried
out successfully (<u>Rec. Trav. Chim.</u>, 1958, <u>77</u>,
854).   Note that no $AlCl_3$ is needed for
Friedel-Crafts alkylation with easily formed
t-alkyl compounds.

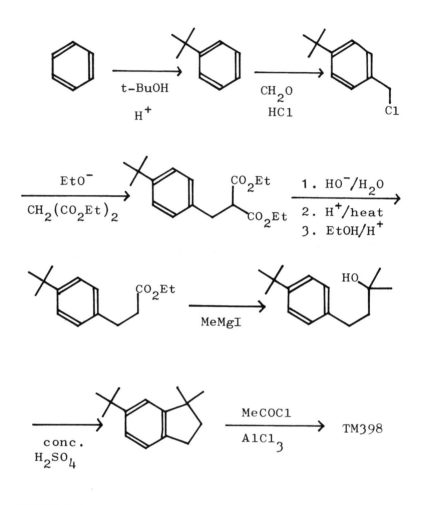

400. <u>Revision Problem 5</u>: This molecule (TM400)
was used by Raphael in his synthesis of the
natural product clovene. How could you make it?

TM400

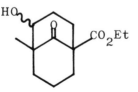

401. <u>Analysis:</u> We want to disconnect any of
the bonds to the common atoms ● to give us a
simple six-membered ring. We can best disconnect bond a as it is part of a 1,3-di oxygen-
ated system:

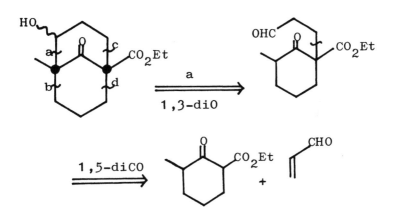

Synthesis: The starting material is similar to TM100, so we shall use the same method:

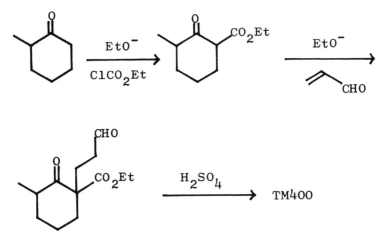

Only isomer A will be formed as the alternative cannot give a stable enolate anion (see
frame 101). This is nearly the synthesis used
by Raphael (Tetrahedron, 1962, 18, 55; Proc.
Chem. Soc., 1963, 239).

---

402. Revision Problem 6: Cascarillic acid
occurs naturally in Euphorbiaceae plants
(spurges). How could you synthesise it?

Cascarillic acid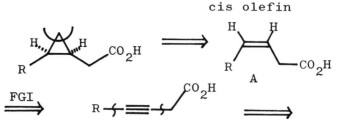

---

403. Analysis: the small ring will dominate
the strategy, and only one disconnection will
make the stereochemistry secure. Writing
R = n-hexyl:

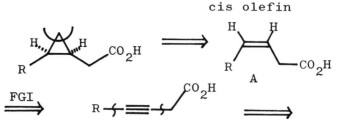

Synthesis: The CO_2H group spells trouble.
We would certainly have to use an ester, but
the α-bromoester is too reactive to use with
an acetylene. Also there is a danger that the
double bond in A will move into conjugation.

We can get round all these problems with an epoxide and then oxidise at the end:

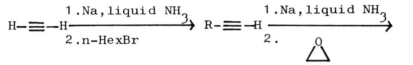

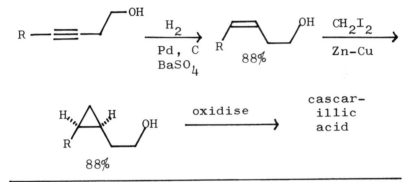

88%

---

404. <u>Review Problem 7</u>: TM404 was needed as an intermediate in a steroid synthesis. How might it be made?

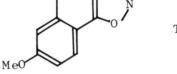

TM404

---

405. <u>Analysis</u>: Taking the heterocyclic part first, we can remove the two heteroatoms as hydroxylamine (the approach of frames 258-261) to give us a 1,3-dicarbonyl compound.

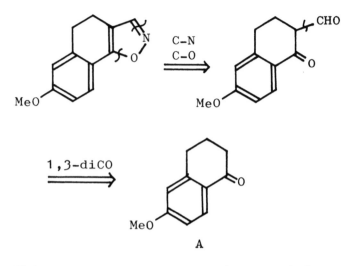

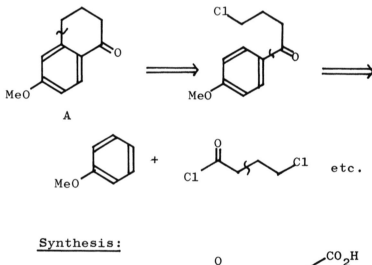

This starting material A (also 224A) is an isomer of the ketone we made in frame 328, and is easy to make using the intramolecular strategy of that frame:

Synthesis:

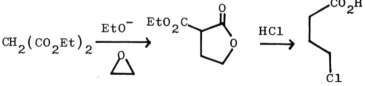

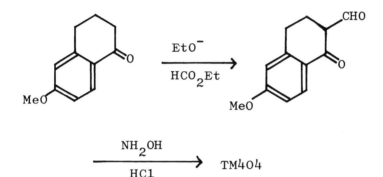

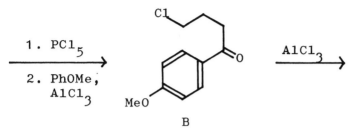

This synthesis was first carried out by Velluz, Angew. Chem., 1960, 72, 725. The lactone can be used instead of the γ-chloro acid, see Org. Synth. Coll., 4, 898. Other approaches to A are outlined in J. Amer. Chem. Soc., 1947, 69, 576, 2936 and it is probable that the reaction actually given here - the cyclisation of B - would give a five membered ring instead (J. Chem. Soc., 1956, 4647). The usual, if illogical way to make A is by reduction of β naphthol methyl ether and oxidation of the product with CrO_3.

406. <u>Revision Problem 8</u>: TM406 is a synthetic
intermediate related to the cannabinoids,
naturally occurring hallucinogenic compounds.
How could it be synthesised?

TM406

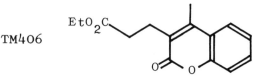

407. <u>Analysis</u>: Another lactone! FGI reveals
the true TM (A). Our normal disconnection <u>a</u>
of an α,β-unsaturated carbonyl compound gives
us the 1,5-dicarbonyl compound (B) and the
ketone (C) clearly derived from phenol. Alter-
natively we could disconnect bond <u>b</u> to the
keto-ester (D) with the further disconnection
shown:

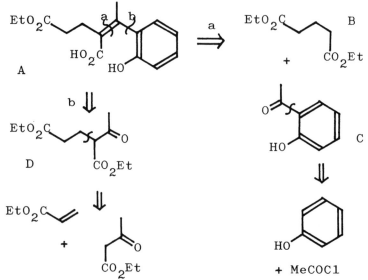

Whichever route we choose, we need <u>ortho</u> substitution. For (C) we can do this via the
Fries rearrangement (PhOH + MeCOCl → PhO.COMe
which rearranges to C with AlCl$_3$) see Norman
p.457-8 or Tedder vol. 2 p.214. However, we
still need the right geometrical isomer of A!
The other route solves this problem because
reaction of D with phenol gives TM406 directly:
ester exchange makes the C-O bond first and
the condensation follows. This is the strategy we discussed in frames 319 ff.

Synthesis: Chem. Ber., 1948, <u>81</u>, 197; J.
Amer. Chem. Soc., 1967, <u>89</u>, 5934: All in two
steps!

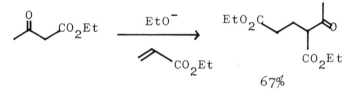

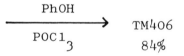

---

408. <u>Revision Problem 9</u>: Suggest a synthesis
for TM408.

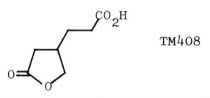

TM408

409. <u>Analysis</u>: This is a γ-lactone and we
spent time considering possible strategies for
these compounds in frames 334-348.   First open
the lactone ring.   This gives us a compound
with 1,4- 1,5- and 1,6-dioxygenation relationships.   I'll follow the 1,6 through.

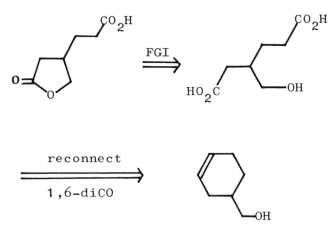

Now we are nearly at a Diels-Alder disconnection so a change in the oxidation state to
aldehyde or ester is needed:

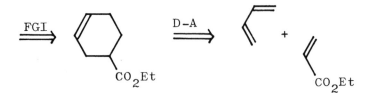

<u>Synthesis</u>: As described in <u>Acta Chem.</u>
<u>Scand. B</u>, 1977, <u>31</u>, 189.

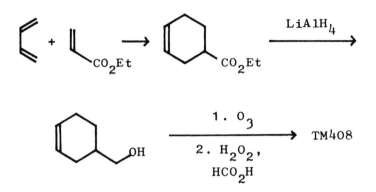

Cyclisation to give the <u>five</u>-membered ring is spontaneous.

---

410. <u>Revision Problem 10</u>: This compound is an intermediate in the synthesis of an alkaloid. Don't worry about the half ether aspect – there is a solution to this which will emerge as you go along.

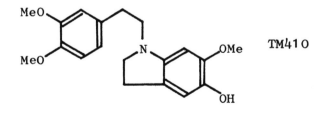

---

411. <u>Analysis</u>: The nitrogen atom is clearly the key to the problem and we can put a carbonyl group on either adjacent CH_2 group. Strategically we make more progress by using the exocyclic CH_2 group first:

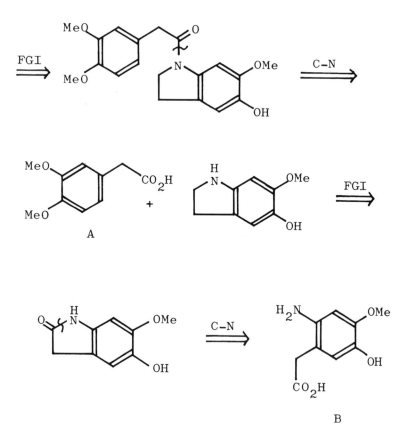

It looks as though we can get B from A (which is used in frame 247) and so the nitro group is the obvious source of the amino group. It will also allow us to hydrolyse one ether specifically by nucleophilic aromatic substitution.

    <u>Synthesis:</u> Starting material as in frame 247, then (E. McDonald and R. Wylie, unpublished work at Cambridge 1976-7):

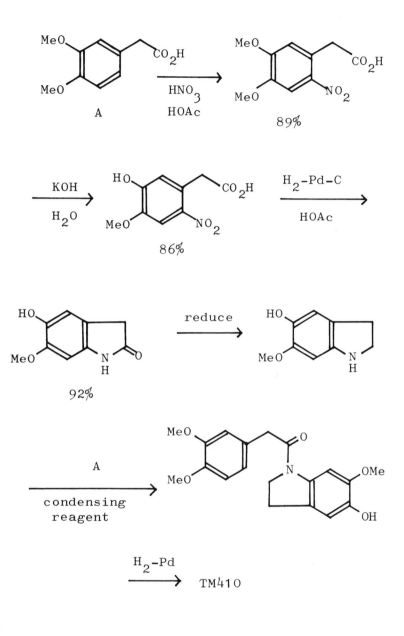

412. These are problems without solutions intended to lead you onto more challenging things.

Strategy Problem 1: "The wrong substitution pattern". Making aromatic compounds m-substituted with two o,p-directing groups is always a problem. What strategies can you suggest? An example (TM412) is the alkyl halide used in the synthesis of some steroids.

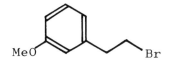

TM412

---

413. Hint - how do you make any m-disubstituted compound? Which of the two side chains is easier to add? How?

Further development: How would you make TM413 using the alkyl bromide you have just made? This molecule is obviously on the way to a steroid and you can read more about it in Helv. Chim. Acta, 1947, 30, 1422 and J. Amer. Chem. Soc., 1942, 64, 974.

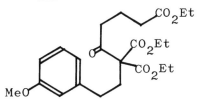

TM413

414. <u>Strategy Problem 2</u>: "The wrong ring sizes". We have said nothing so far about seven membered rings. The common arrangement in natural products is a fused seven-five ring system as in the molecules below. They are often synthesised from six-six fused systems because we understand them so much better. What particular six-six fused molecules might you use to make these molecules:

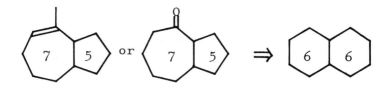

You can find some answers in <u>J. Amer. Chem. Soc.</u>, 1966, <u>88</u>, 4113; 1969, <u>91</u>, 6473; 1971, <u>93</u>, 1746; <u>Org. Synth. Coll.</u>, <u>5</u>, 277.

---

415. <u>Strategy Problem 3</u>: "Too large a ring". Medium rings are also tricky problems and one way to make them is to cleave a bond in a bicyclic compound, e.g.

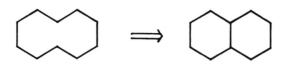

What kind of reactions could be used to implement this strategy? How in particular might you make these compounds:

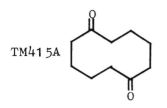

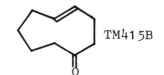

note trans olefin

You will find some solutions in: J. Amer. Chem Soc., 1963, <u>85</u>, 362; 1964, <u>86</u>, 485, Ber., 1933, <u>66</u>, 563. Tetrahedron Letters, 1976, 4409.

---

416. <u>Strategy Problem 4</u>: "The wrong polarity". You have seen how important it is to have reagents corresponding to as many synthons as possible. One we haven't mentioned is the acyl anion R-CO$^-$ for which we had one reagent if R=Me (frame 145 if you've forgotten!). Can you devise any more general reagents for this synthon?   Ideally we should like to make ketones this way:

$$R^1CO^- + R^2{-}Hal \longrightarrow R^1CO.R^2$$

---

417.   This is a very challenging problem indeed.   You may find some solutions by setting up two substituents on the carbon atom. For example, how about a substituted Wittig reagent:

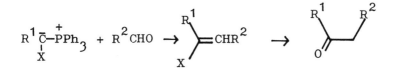

What could X be for this sequence to be feas-
ible? You will find an account of some modern
solutions in <u>Tetrahedron</u>, 1976, <u>32</u>, 1943 and
<u>Synthesis</u>, 1977, 357.

---

418. <u>Strategy Problem 6</u>: A labelled compound
for biosynthetic studies. Mevalonic acid
(TM418) is an intermediate in the biosynthesis
of terpenes and steroids (Tedder, volume 4,
p.217 ff). To study exactly what happens to
each carbon atom during its transformation into, say, limonene (418A), we need separate
s amples of mevalonic acid labelled with ^{14}C
in each carbon atom in the molecule. This
turns our normal strategy on its head since
we must now look for one carbon disconnections.
You can use reagents like $^{14}CH_3I$, Na^{14}CN, and
$^{14}CH_3CO_2H$. See if you can find approaches to
some of the labelled compounds.

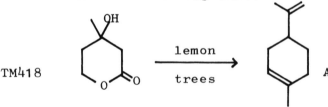

TM418

Some solutions can be found in <u>Tetrahedron</u>,
1959, <u>5</u>, 311, <u>Chem. Abs.</u>, 1966, <u>65</u>, 614, <u>J.
Amer. Chem. Soc.</u>, 1975, <u>97</u>, 4144 and see J.W.
and R.H. Cornforth in "Natural Substances
formed Biologically from Mevalonic Acid", ed.,
T.W. Goodwin, Academic Press, 1970, p.5 where
all the different routes are explained.

419. <u>Strategy Problem 7</u>: Synthesis of a single enantiomer. Many compounds such as pharmaceuticals, flavourings, and insect control chemicals must not only have the right relative stereochemistry but must be optically active too if they are to be of any use. Consider the strategy of synthesising one enantiomer:

(a) should you start with an optically active compound?

or (b) should you resolve at some point?
if so(c) which is the best point for a resolution?
and (d) what do you do with the unwanted enantiomer?

You can take some examples from this list for your consideration:

Target molecules in frames: 133, 135, 152 216, 242, 249, 251, 377.

Review Problems	8	(frame 110)
	26	(frame 269)
	27	(frame 271)
	32	(frame 303)
Revision Problems	6	(frame 402)
	9	(frame 408)
γ-Lactones		(frames 342-348)
Baclofen		(frames 349-354)
Chrysanthemic acid		(frames 355-370)

420. Briefly, the answers to the questions
are:

    (a) Yes, if one is available - it usually
isn't.

    (b) You probably have to.

    (c) As early as possible.

    (d) Recycle it or use it to resolve some-
thing else.

The question of stereochemical control has
been a theme running throughout the programme
and as you progress to more complicated molecules it becomes more important. This is very
clear from many of the syntheses described in
Fleming.

421. This cyclopentadione is needed to provide
ring D in some steroid synthesis. Unlike the
corresponding six-membered ring compound it is
difficult to make. Can you suggest any
solutions?

Published Solutions: Bull. Soc. Chim.
France, 1955, 1036; 1965, 645; Org. Synth.,
1967, 47, 83; J. Org. Chem., 1967, 32, 1236;
Angew. Chem., 1967, 79, 97, 378; Chem. Ber.,
1967, 100, 2973; 1969, 102, 3238.

422. Cis-Jasmone (TM422) is an important
ingredient in many perfumes. There are several
obvious disconnections and it may help you to
know that cyclisation of the diketone 422A
does indeed selectively give cis-jasmone.

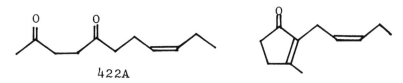

422A

TM422

Published Solutions: J. Chem. Soc., 1969
(C), 1016, 1024; Chem. Comm., 1972, 529;
Tetrahedron Letters, 1972, 1233; 1976, 4867;
1971, 1569, 2575; 1973, 3267, 3271, 3275;
1974, 3883, 1237, 1387, 4223.

J. Amer. Chem. Soc., 1964, 86, 935, 936; 1970, 92, 7428; 1971; 93, 5309, 3091; 1972, 94, 8641; 1973, 95, 4446, 4763. Canad. J. Chem., 1972, 50, 2718. J. Org. Chem., 1972, 37, 341, 2363; 1966, 31, 977; 1971, 36, 2021. Chem. Lett., 1972, 793; 1973, 713.

Fourteen syntheses of cis-jasmone are given in chart form in 'Natural Products Chemistry' ed. K. Nakanishi et. al., Academic Press, New York, 1975, Vol.2, p.21.

423. Juvabione is a substance produced by some conifers in imitation of a hormone in an insect pest. It may be a kind of natural control of the pest as it prevents it reaching maturity.

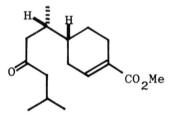

Juvabione

Published Solutions: Tetrahedron Letters, 1967, 2515, 4677; 1969, 351; Tetrahedron, 1968, 24, 3127; Chem. Comm., 1968, 1057; Canad. J. Chem., 1968, 46, 1467; J. Amer. Chem. Soc., 1970, 92, 336.

424. Grandisol, with a four-membered ring, is
another insect hormone, the male sex hormone
of the boll weevil to be precise.   It may also
be useful as a highly specific pest control.
How might it be made?

Grandisol

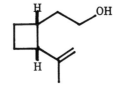

Published Solutions: Science, 1969, 166,
1010; J. Amer. Chem. Soc., 1970, 92, 425; 1974,
96, 5268, 5270, 5272; 1976, 98, 4594; J. Org.
Chem., 1972, 37, 1854 .